M000286834

NUMBER POWER

A REAL WORLD APPROACH TO MATH

Measurement

 **Education**

Bothell, WA • Chicago, IL • Columbus, OH • New York, NY

www.mheonline.com

Copyright © 2011 and 1996 by The McGraw-Hill Companies, Inc.

Cover 5 127 Jeff Spielman/Photographer's Choice RF/Getty Images.

Send all inquiries to:
Contemporary/McGraw-Hill
130 E. Randolph, Suite 400
Chicago, IL 60601

ISBN: 978-0-07-659234-0
MHID: 0-07-659234-0

Printed in the United States of America.

1 2 3 4 5 6 7 8 9 RHR 15 14 13 12 11

TABLE OF CONTENTS

Capacity and Volume

Time and Velocity

USING NUMBER POWER

Posttest

Answer Key

Formulas and Measurement Units

Glossary

Index

TO THE STUDENT

This Number Power book is designed to help you build and use the basic math skills you need to handle measurement situations found in many work settings and in your everyday experiences. This practice also will help you improve your basic math ability.

Each section of this book provides settings where measurement skills are used. There are step-by-step examples and plenty of practice exercises to develop and reinforce your understanding and skill with measurements. Throughout the book you have opportunities to apply your math skills in common, real measurement situations. Both customary units and metric units are presented throughout the book. Also, whole numbers, fractions, and decimals are used throughout the book.

Learning how and when to use a calculator is an important skill to develop. In real life, problem solving often involves the smart use of a calculator—especially when working with large numbers. You need to know when an answer on a calculator makes sense, so be sure to have an estimated answer in mind. Remember that a calculator can help you only if you set up the problem and use the calculator correctly.

At the back of this book are an index, a glossary, formulas, and an answer key for all the exercises. The index, glossary, and formulas are useful tools for quick review and helpful explanation. Using the Answer Key after you have worked a lesson will help you check your progress.

The measurement skills you develop and practice through this book will be helpful in your work, at home, at school, and in other everyday situations.

Pretest

Math and Measurement Skills Inventory

1. $\begin{array}{r} 2{,}438 \\ +\ 6{,}351 \\ \hline \end{array}$

2. $\begin{array}{r} 306{,}935 \\ +\ 292{,}042 \\ \hline \end{array}$

3. $28 + 46 + 37 =$

4. $5{,}643 + 975 + 1{,}605 =$

5. $4{,}738 + 921 + 16 + 1{,}204 =$

6. $\begin{array}{r} 387 \\ -\ 142 \\ \hline \end{array}$

7. $\begin{array}{r} 4{,}052 \\ -\ 1{,}021 \\ \hline \end{array}$

8. $27{,}005 - 18{,}927 =$

9. $3{,}049 - 956 =$

10. $\begin{array}{r} 427{,}361 \\ -\ 284{,}196 \\ \hline \end{array}$

11. $\begin{array}{r} 53 \\ \times\ 3 \\ \hline \end{array}$

12. $\begin{array}{r} 46 \\ \times\ 8 \\ \hline \end{array}$

13. $\begin{array}{r} 6{,}102 \\ \times\ \quad 4 \\ \hline \end{array}$

14. $27 \times 35 =$

15. $\begin{array}{r} 607 \\ \times\ 18 \\ \hline \end{array}$

16. 789×65

17. $\begin{array}{r} 18{,}475 \\ \times\quad 609 \\ \hline \end{array}$

18. $619 \times 200 =$

19. $\begin{array}{r} 1{,}000 \\ \times\quad 347 \\ \hline \end{array}$

20. $5\overline{)175}$

21. $7\overline{)2{,}940}$

22. $8\overline{)4{,}630}$

23. $28\overline{)4{,}936}$

24. $4{,}780 \div 40 =$

25. $4{,}670 \div 1{,}000 =$

26. Reduce $\frac{18}{24}$.

27. $\frac{1}{4} + \frac{1}{4} =$

28. $\frac{3}{8} + \frac{1}{4} =$

29. $\begin{array}{r} 2\frac{1}{2} \\ + 3\frac{1}{3} \\ \hline \end{array}$

30. $\begin{array}{r} 1\frac{3}{4} \\ + 4\frac{1}{2} \\ \hline \end{array}$

31. $\frac{4}{5} - \frac{1}{5} =$

32. $\frac{2}{3} - \frac{1}{2} =$

33. $\begin{array}{r} 6 \\ - 3\frac{2}{3} \\ \hline \end{array}$

34. $\begin{array}{r} 4\frac{1}{2} \\ - 1\frac{3}{5} \\ \hline \end{array}$

35. $\frac{1}{2} \times \frac{7}{10} =$

36. $4\frac{1}{2} \times 3\frac{1}{3} =$

37. $2\frac{1}{4} \times \frac{2}{3} =$

38. $\frac{4}{5} \div \frac{1}{2} =$

39. $2\frac{1}{2} \div \frac{1}{2} =$

40. $1\frac{1}{4} \div 2 =$

41. $6\frac{1}{2} \div 1\frac{1}{4} =$

42.   2.6
         $+\ 4.8$

43. $2.73 + 4.2 + 6.0 =$

44.   42.68
         $+\ 39.842$

45.   6.95
         $-\ 3.74$

46. $206.6 - 48.75 =$

47.   578.42
         $-\ 229.56$

48.   206
         $\times\ 4.05$

49.   37.6
         $\times\ 0.01$

50.   42.56
         $\times\ \ 100$

51. $496.35 \times 2.6 =$

52. $2.2\overline{)46.2}$

53. $0.01\overline{)3.5}$

54. $58\overline{)997.6}$

55. $84.75 \div 100 =$

56. The Smith family spent $152 for electricity in April, $146 in May, $158 in June, and $166 in July. What was their average monthly bill for electricity for the four months?

57. Jai prepares tax returns. His client earns $820 every two weeks. Jai needs to report his client's earnings for one year. How much does the client earn in one year?

58. Jim's original weight was 206 pounds. If he lost 28 pounds and then gained 12 pounds, what is his current weight?

59. Jeni teaches a sewing class of 8 students. She needs to give each student $2\frac{3}{4}$ yards of fabric for a project. How much fabric does she need?

60. Working as a painter, Josie began a job with four and a half gallons of paint. At the end of the job, she had $1\frac{3}{4}$ gallons left. How much paint did she use for the job?

61. Teresa bought a 30-yard spool of lace trim. She is decorating cloths, and each cloth takes $1\frac{1}{2}$ yards of lace. How many cloths can she decorate with the 30-yard spool?

62. Juan is an office manager. He had $247.58 in a checking account for office supplies. He wrote a check for $36.95 and then deposited $62.46. What is the new balance?

63. Tami bought ground beef for $2.59 per pound. If she paid $13.52, how many pounds of ground beef (to the nearest tenth) did she buy?

SKILLS INVENTORY CHART

If you got fewer right than you should have, or if you cannot tell why your answers are not correct, you can review sections in the other Number Power books. There you will find explanations and more practice problems for each skill in this inventory.

Rework any problems you missed in the Skills Inventory. Then, when you are satisfied with your results, begin work in this Measurement book.

Problem Numbers	Other *Number Power* Books
1–25, 56–58	*Number Power: Addition, Subtraction, Multiplication, and Division*
26–55, 59–63	*Number Power: Fractions, Decimals, and Percents*

BUILDING
NUMBER
POWER

MEASUREMENTS

When Do We Use Them?

We use measurements when we describe people, places, and things. For example, a driver's license describes a person's height, weight, and age; and a building plan describes lengths, heights, angles, areas, and volumes.

We use measurements when we want to compare two things. For example, you can tell if one liquid is hotter than another by measuring the temperature of each one. Or, you can tell if one motorboat is faster than another by measuring the top velocity of each one.

We use measurements when we want to find differences between things. For example, to find how much gasoline you used on a trip, you could find the difference between the number of gallons you started with and the number of gallons you ended with. Or, to find the length of a car trip, you could subtract the start-of-trip odometer reading from the end-of-trip odometer reading.

All measurements have two parts, a number and a unit. The number may be a whole number, a fraction, or a decimal. So, when you use measurements you will be practicing your skills at adding, subtracting, multiplying, and dividing with different types of numbers.

In this book, you will learn about many different kinds of units, including units for weight, length, temperature, and other physical characteristics of objects. Some units, like ounces, pounds, feet, quarts, and gallons, are **Customary Units of Measurement** (sometimes called **Imperial units**). Other units, like grams, meters, and liters, are **Metric Units of Measurement** (sometimes called **International units**).

> As you use measurements, you will develop and review your math skills involving whole numbers, decimals, and fractions. Measurements are important for almost all jobs, for many daily activities, and for understanding and solving problems. In these lessons, you will investigate tools and scales for finding measurements. Also, you will apply those tools and scales using different kinds of units.

Customary and Metric Units

Does your city or town have a water tower? If so, it might look like the tower at the right. To describe this water tower, we might say that it is 115 feet tall, weighs 16,000 pounds (when empty), and can hold 25,000 gallons of water.

In this description, 115 feet is a measurement of *length*, 16,000 pounds is a measurement of *weight*, and 25,000 gallons is a measurement of *capacity*. Here is how some of the customary units for length, weight, and capacity are related to each other.

Length	Weight	Capacity
1 foot (ft) = 12 inches (in.)	1 pound (lb) = 16 ounces (oz)	1 cup (c) = 8 fluid ounces (fl oz)
1 yard (yd) = 36 in.	1 ton (T) = 2,000 lb	1 pint (pt) = 2 c
1 yd = 3 ft		1 quart (qt) = 2 pt
1 mile (mi) = 5,280 ft		1 gallon (gal) = 4 qt
1 mi = 1,760 yd		

Study the chart above. What is the abbreviation for each unit?

1. yards _____

2. tons _____

3. inches _____

4. pints _____

5. miles _____

6. fluid ounces _____

7. pounds _____

8. gallons _____

9. quarts _____

Fill in each blank with the correct equivalent measurement.

10. 1 c = _____ fl oz

11. 1 ft = _____ in.

12. 1 ton = _____ lb

13. 1 lb = _____ oz

14. 1 gal = _____ qt

15. 4 ft = _____ in.

16. 1 pt = _____ c

17. 1 mi = _____ yd

18. 3 gal = _____ qt

Name three things that are sold, measured, or packaged in the given unit.

19. miles _____

20. gallons _____

21. ounces _____

22. pounds _____

In metric units, the basic unit of length is the **meter**, the basic unit of weight is the **gram**, and the basic unit of capacity is the **liter**.

Most track-and-field races are measured in meters. A meter is a little longer than a yard. Some medicines are sold in grams. A gram is approximately the weight of a vitamin tablet. Many large soft-drink containers are measured in liters. A liter is a little more than a quart.

The metric system uses two important prefixes at the beginning of a unit. They are "milli-," which means "one thousandth of," and "kilo-," which means "one thousand times." Here is how the metric units are related to each other.

Length	Weight	Capacity
1 meter (m) = 1,000 millimeters (mm)	1 gram (g) = 1,000 milligrams (mg)	1 liter (L) = 1,000 milliliters (mL)
1 kilometer (km) = 1,000 m	1 kilogram (kg) = 1,000 g	1 kiloliter (kL) = 1,000 L

Study the chart above. What is the abbreviation for each unit?

23. milliliter _____

24. kilogram _____

25. meter _____

26. kilometer _____

27. liter _____

28. milligram _____

29. gram _____

30. millimeter _____

31. kiloliter _____

Fill in each blank with the correct equivalent measurement.

32. 1 kg = _____ g

33. 1 km = _____ m

34. 1 m = _____ mm

35. 1 g = _____ kg

36. 1 kL = _____ L

37. 1 L = _____ mL

Name two things that are sold, measured, or packaged in the given unit.

38. meters _____

39. kilometers _____

40. grams _____

41. liters _____

Reading Rulers and Straight-Line Scales

A ruler is an example of a simple scale. Here are the steps involved in reading a measurement from a ruler.

EXAMPLE 1   How long is the pencil?

STEP 1   Place the ruler so the zero mark on the ruler lines up with one end of the pencil.

STEP 2   The units on the ruler are inches. Between which two whole inches is the length of the pencil?

3 and 4

STEP 3   Count the number of marks between each inch mark. What does each mark represent?

There are 16 marks between each pair of inch marks. Each mark represents $\frac{1}{16}$ inch.

STEP 4   How many $\frac{1}{16}$-inch marks does the pencil extend beyond 3 inches?

Three of them

ANSWER: The pencil is about $3\frac{3}{16}$ inches long.

You can read thermometers and many other straight-line scales the same way you read the scale on a ruler.

EXAMPLE 2   Mark this outdoor thermometer so it shows 73°F. ("°F" stands for "degrees Fahrenheit.")

90° 80° 70° 60° 50° 40° 30° °F

STEP 1   Count the spaces between 70° and 80°. What does each space represent?

There are 5 spaces between 70 and 80. Each space represents 2°F.

STEP 2   How many spaces beyond 70° are needed to show 73°F?

$1\frac{1}{2}$ spaces

ANSWER: 90° 80° 70° 60° 50° 40° 30° °F
73°

Write the measurement shown on each Fahrenheit or Celsius scale.

1.

_____

2.

_____

3.

_____

4.

_____

Mark the given measurement on each scale.

5.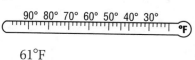

$3\frac{5}{8}$ inches

6.

$\frac{7}{8}$ inches

7.

61°F

8.

74°F

9.

$5\frac{5}{16}$ inches

10.

$5\frac{1}{8}$ inches

11.

$2\frac{1}{4}$ inches

12.

19°C

13.

32°C

14.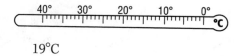

90.5°F

Reading Arcs and Dials

If you have a gas, water, or electric meter at your home, it may have circular dials. Here is an example of how to read a four-dial meter. (Notice that the word *meter*, as used here, is not a unit of length. Its meaning is close to that of a "parking meter.")

EXAMPLE 1 Electricity is usually measured in kilowatt-hours (kWh). Meter-readers sometimes need to read four-dial meters such as this. What is the reading for this four-dial electric meter?

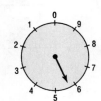 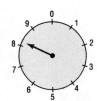 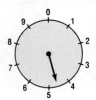 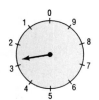

STEP 1 Start at the left. The pointer is between which two whole numbers? *4 and 5*

STEP 2 Move to the next dial. The pointer is between which two whole numbers? *5 and 6*

STEP 3 For each of the next two dials, the pointer is between which two whole numbers? *8 and 9, 2 and 3*

ANSWER: The reading of the four-dial meter is **4,583 kWh**.
(The meter-reader usually rounds up for the last dial.)

Express each four-meter reading in kilowatt-hours.

1.
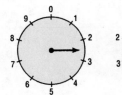

Reading: _____

2.
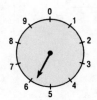

Reading: _____

3.

Reading: _____

EXAMPLE 2 **Mark a two-dial meter so it shows a reading of 84 kWh.**

STEP 1 Start at the left. Find 8 and 9 on the dial. Draw a pointer between them.

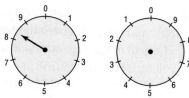

STEP 2 Now go to the dial at the right. Find 3 and 4 on the dial. Draw a pointer between them.

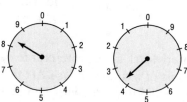

ANSWER: The two-dial meter shows a reading of **84 kWh**.

Mark each four-dial meter with the given reading.

4. 2,735 kWh

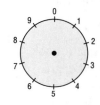 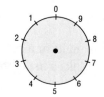 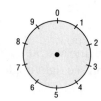

5. 4,598 kWh

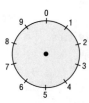 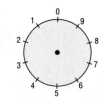 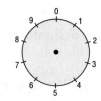

6. 3,009 kWh

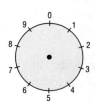 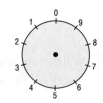 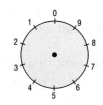

Mark each five-dial meter with the given reading.

7. 42,351 kWh

  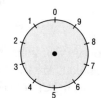

8. 30,037 kWh

  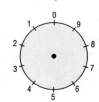

Measuring Around the House

A carpenter's level may use a protractor and a glass tube with an air bubble and water to check the walls and floors of a building. If a wall or the side of a door is straight up-and-down, or vertical, it is called **plumb**. If a floor, ceiling, or doortop is perfectly horizontal, it is called **level**. The protractor is used to measure the angle formed by the wall or floor.

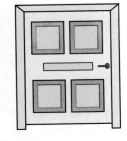
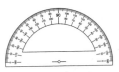

1. Use a carpenter's level and a protractor to describe the angle formed by the walls, floors, and doors of your own house. Are the walls plumb? Are the floors level? If not, what angles do they form with a horizontal line? Many houses "settle" after a few years, when the foundation compresses the earth below it, or the wood in the framing of the house shrinks slightly.

2. Check the bottom of a bathtub, a driveway, or the floor in a basement near a floor drain. What angles do these surfaces form with a horizontal line? Should these surfaces be level?

3. A horizontal line meets a vertical line at a 90-degree angle, or a right angle. Look around your house. Where do you see right angles? Make a list of the 90-degree angles in your house.

The weight of an object is how heavy or light it is. Here are some activities involving weight.

4. Using a bathroom scale, weigh yourself with an overcoat and all your clothes on. Then weigh yourself wearing fewer clothes. How can you find the weight of the clothing you removed without putting that pile of clothes on the scale? What is the weight of the clothing?

5. Use a bathroom scale to weigh a pet or a small child: Weigh yourself holding the pet or child, then weigh yourself alone. What math operation can you use to find the weight of the pet or child? What is that weight?

6. If you have a small scale, weigh some of your letters. How much do they weigh? For a first-class letter, what is the most it can weigh without needing more than a first-class stamp?

7. Before you cook a piece of meat, poultry, or fish, use a kitchen scale to weigh it. Then weigh the food after you cook it. Are the two weights the same? If the weights are different, what do you think happened?

The temperature of a person or an object is a measure of how much heat it contains or is giving off. Here are some activities involving temperature.

8. The "normal" body temperature for people is 98.6°F. This means that most people, when they are healthy, have a temperature close to 98.6°F. Use a personal thermometer to take your temperature before and after exercising. If the two temperatures are different, what do you think caused the difference?

9. Using a thermometer that has a range from at least 40°F to 200°F, find the temperature of a "hot shower," a "cold shower," a "hot bath," and a "cool bath." What is the most comfortable water temperature for you when you shower or bathe? (CAUTION: Do not use a "fever" thermometer for this kind of activity. A temperature above 106°F may destroy that kind of thermometer.)

10. A thermostat helps regulate the temperature of your home. Use a thermometer to find the temperature of several rooms in your home, including places high in the house, low in the house, and far from the thermostat. What is the range in temperature readings? Are they close to the thermostat reading?

11. In your city or town, what is a comfortable setting for a thermostat from December to March? What is a comfortable setting for a thermostat from July to September? If these two settings are different, give an explanation why this may be so.

The perimeter of a figure is the distance around the figure. The area of a figure is the size of the region enclosed by the figure. Here are some activities involving perimeter and area.

12. The figure below at the left shows the floor of a large room. The floor is covered in one-foot by one-foot square tiles. What is the area of the floor? What is the perimeter of the floor? How are the area and perimeter related to the length and width of the floor?

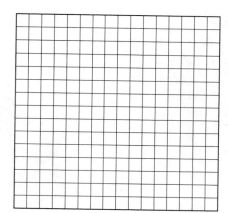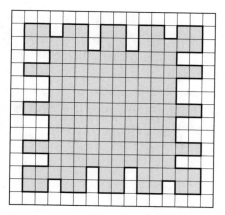

13. The figure above at the right shows the same room with an unusual rug covering part of the floor. What is the area of the rug? What is the perimeter of the rug?

14. In the two figures above, the rug covers part of the floor, so the rug is not as large as the floor. Think about your measurements in exercises 12 and 13. Can part of the floor ever have a greater area than the whole floor? Can part of the floor ever have a greater perimeter than the whole floor?

Estimating Length in Meters

The basic metric units for length are **millimeter, centimeter, meter,** and **kilometer.** Here are the relationships among these units.

1 meter (m) = 100 centimeters (cm)	1 cm = 10 mm
1 m = 1,000 millimeters (mm)	1 km = 1,000 m

width of your
little finger: 1 cm

width of a needle: 1 mm

height of a
six-year-old: 1 m

distance around a
large museum: 1 km

EXAMPLE 1 Janis works for a shipping company. She wants to know the thickness of a telephone directory so she can decide how many directories will fit in a box. She uses a meter stick, a ruler one meter long, to measure the thickness. What is the thickness of the directory?

STEP 1 Place the meter stick so one end is even with one of the covers.

STEP 2 The numbers on the meter stick are centimeters. Between which two whole centimeters is the other cover?   *11 and 12*

STEP 3 Read the number on the meter stick.   *about 11.9 cm*

ANSWER: The thickness of the directory is **about 11.9 cm**.

Name three things that might have each given length.

1. 1 millimeter

_____

2. 1 centimeter

_____

3. 1 meter

_____

4. 1 kilometer

_____

Write the measurement shown by each arrow.

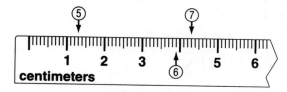

5. _____   **6.** _____   **7.** _____

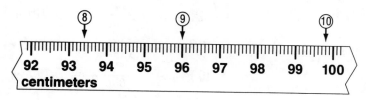

8. _____   **9.** _____   **10.** _____

Use an arrow and a number to mark each measurement on the metric ruler.

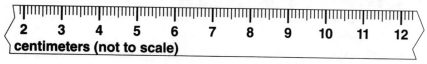

11. 5 cm   **12.** 8 cm 6 mm   **13.** 4 cm 8 mm

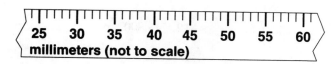

14. 28 mm   **15.** 32 mm   **16.** 51 mm

Metric Lengths and the Decimal Point

Metric units use a decimal to represent a part of a unit.

1 m = 100 cm	so	1 cm = 0.01 m
1 m = 1,000 mm	so	1 mm = 0.001 m
1 km = 1,000 m	so	1 m = 0.001 km

EXAMPLE 1 The widths of some metric machine parts are 19 millimeters, 48 millimeters, and 197 millimeters. How can you express each measurement in centimeters?

STEP 1 Write the relationship between mm and cm.

$$10 \ mm = 1 \ cm$$
$$1 \ mm = 0.1 \ cm$$

STEP 2 To change from mm to cm, divide by 10. This is the same as moving the decimal point one place to the left.

$$19 \ mm = \frac{19}{10} \ cm = 1.9 \ cm$$
$$48 \ mm = \frac{48}{10} \ cm = 4.8 \ cm$$
$$197 \ mm = \frac{197}{10} \ cm = 19.7 \ cm$$

ANSWER: The measurements are **1.9 centimeters**, **4.8 centimeters**, and **19.7 centimeters**.

EXAMPLE 2 Express the measurements from Example 1 in meters.

STEP 1 Write the relationship between millimeters and meters.

$$1,000 \ mm = 1 \ m$$
$$1 \ mm = 0.001 \ m$$

STEP 2 To change from mm to m, divide by 1,000. This is the same as moving the decimal point three places to the left.

$$19 \ mm = \frac{19}{1,000} \ m = 0.019 \ m$$
$$48 \ mm = \frac{48}{1,000} \ m = 0.048 \ m$$
$$197 \ mm = \frac{197}{1,000} \ m = 0.197 \ m$$

ANSWER: The measurements are **0.019 meters**, **0.048 meters**, and **0.197 meters**.

Rewrite each measurement so it uses the given unit.

1. 2,000 m = _____ km

2. 1,000,000 cm = _____ m

3. 750,000 cm = _____ km

4. 300 mm = _____ cm

5. 600 m = _____ km

6. 1,000,000 mm = _____ m

7. 2,200 m = _____ km

8. 625 cm = _____ m

9. 1,475 mm = _____ cm

10. 8,700 mm = _____ m

11. 4,900 m = _____ km

12. 18,800 cm = _____ km

EXAMPLE 3 The distance between two towns is 15.83 kilometers. Explain how to change 15.83 kilometers to meters.

STEP 1 Write the relationship between km and m.

$1 \, km = 1000 \, m$

STEP 2 To change km to m, multiply by 1,000. This is the same as moving the decimal point three places to the right.

$15.83 \, km = (15.83)(1,000) \, m$
$= 15,830 \, m$

ANSWER: 15.83 km = **15,830 m**

Rewrite each measurement so it uses the given unit.

13. 3 m = _____ cm

14. 529 cm = _____ mm

15. 27 m = _____ cm

16. 21 km = _____ m

17. 7 km = _____ cm

18. 12 m = _____ mm

19. 2.25 m = _____ mm

20. 3.2 cm = _____ mm

21. 40.5 km = _____ m

22. 1.6 km = _____ m

23. 12.75 cm = _____ mm

24. 5.2 km = _____ m

25. 8.2 km = _____ cm

26. 3.6 m = _____ cm

Solve each problem.

27. Matt, a cabinet maker, needs a board 23.5 centimeters wide for a bookshelf he is building. He uses a ruler marked in millimeters to check the width of the board. What should be the width of the board in millimeters?

28. Tim, a delivery person, drove 21.68 kilometers. How many meters did he drive?

29. Jana makes picture frames. Each day she must record how many meters of a certain type of framing she uses. One day she used 580 centimeters of the framing. How many meters should she record?

30. Georgia is a swim coach. Each week she records the number of kilometers her team swims. Last week the team swam 3,500 meters. How many kilometers did the team swim?

Operations with Metric Measurements

To add, subtract, multiply, or divide with metric measurements, express each measurement using just one unit. Then use paper and pencil or a calculator to do the operation.

EXAMPLE 1   A plumber has a length of pipe 3 meters and 15 centimeters long. He cuts it into 4 equal pieces. What is the length of each piece?

STEP 1   Write the measurement using a single unit.

$$3\,m + 15\,cm = 3\,m + \frac{15}{100}\,m$$
$$= 3\,m + .15\,m$$
$$= 3.15\,m$$

STEP 2   Divide 3.15 by 4.

$$3.15 \div 4 = 0.7875$$

ANSWER: The length of each piece of pipe is **0.7875 m**. That is the same as 78.75 cm.

EXAMPLE 2   Add: 3.75 meters plus 850 centimeters.

STEP 1   Write 850 cm in meters.

$$850\,cm = 8.5\,m$$

STEP 2   Add the numbers.

$$3.75 + 8.5 = 12.25$$

ANSWER: The total is **12.25 m**.

For each operation, use a single unit to find the answer.

1. $15 \times (10\,m + 350\,cm)$

2. $6 \times (2\,km + 135\,m)$

3. $(4\,cm + 18\,mm) \div 5$

4. $450\,mm - 2\,cm$

5. $3\,km + 300\,m + 300\,mm$

6. $(4\,m + 37\,cm) + (17\,m + 250\,cm)$

Solve each problem.

7. A fence construction crew measures a rectangular field. One crew member reports the length as 0.045 kilometers. Another member reports the width as 25 meters. How many kilometers of fence are needed to go all around the field?

8. To build the fence in problem 7, the crew uses 15 equal sections for one of the 0.045-kilometer sides. How many meters long is each section of fencing?

Estimating Length Using Customary Units

The customary units of length probably are familiar to you: **inches, feet, yards,** and **miles.** Here are the relationships among these units.

12 inches (in.)	=	1 foot (ft)	5,280 ft	=	1 mile (mi)
3 ft	=	1 yard (yd)	1,760 yd	=	1 mi
36 in.	=	1 yd			

height of a
paperclip: 1 in.

height of this
book: just under 1 ft

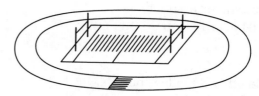

four times around
a track: 1 mi

Name four things that might be measured in each unit.

1. inches

_____

2. feet

_____

3. yards

_____

4. miles

_____

Circle the item you think is longer.

5. a pencil or a jump rope

6. a hammer or a nail file

7. an extension cord or a yardstick

8. a key or a screwdriver

Circle the more appropriate measurement for the length of each object.

9. height of a 6-year-old child
   38 in. or 38 ft

10. distance of a "sprint" race
    100 yd or 100 mi

11. length of a window frame
    42 ft or 42 in.

12. length of a bed sheet
    86 yd or 86 in.

13. height of a giraffe
    18 yd or 18 ft

14. height of a flag pole
    60 ft or 60 yd

15. distance from Chicago to St. Louis
    325 mi or 325 ft

16. length of a garden hose
    50 ft or 50 yd

17. length of a swimming pool
    25 yd or 25 in.

18. distance from Portland, OR, to L.A.
    410 yd or 410 mi

In each group, underline the shortest object.

19. a pair of pliers
    a pair of tongs
    a spatula

20. a man
    a child
    a pony

21. a postcard
    a business envelope
    a roll of adhesive tape

22. a car
    a tricycle
    a truck

23. a chair
    a curtain for a patio door
    a dish towel

EXAMPLE Carpenters often measure with a yardstick, which is a ruler 36 inches, or 1 yard, long. A carpenter measures the distance between two pencil marks on a board, as shown in the diagram. What is the distance between the two pencil marks?

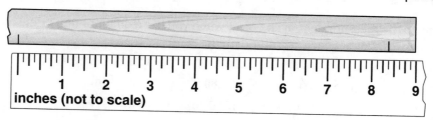

inches (not to scale)

STEP 1 Place the yardstick so one of the pencil marks is at the zero point.

STEP 2 Between which two whole inches is the second pencil mark?

8 and 9

STEP 3 Look at the number of tick marks on the yardstick between 8 and 9. What does each tick mark represent?

There are 8 tick marks, so each tick mark represents $\frac{1}{8}$ inch.

STEP 4 Read the distance between the pencil marks.

8 inches and $\frac{3}{8}$ of an inch

ANSWER: The distance between the two pencil marks is $8\frac{3}{8}$ **inches**.

Write the measurement shown by each arrow.

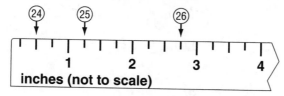

24. _____   25. _____   26. _____

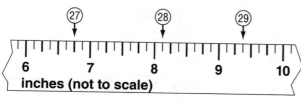

27. _____   28. _____   29. _____

Label the yardstick to show each length.

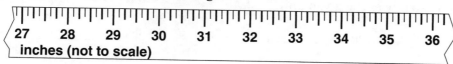

30. $31\frac{5}{8}$ in.   32. $32\frac{1}{2}$ in.   34. $35\frac{7}{8}$ in.

31. 34 in.   33. $33\frac{1}{4}$ in.   35. $35\frac{1}{8}$ in.

Changing Among Inches, Feet, Yards, and Miles

What happens to the measurement of an object when you use a different unit? Here are the relationships between inches, feet, yards, and miles.

12 inches (in.)	=	1 foot (ft)		5,280 ft	=	1 mile (mi)
3 ft	=	1 yard (yd)		1,760 yd	=	1 mi
36 in.	=	1 yd				

EXAMPLE 1 Toni is a carpenter. She needs baseboard molding to put around a rectangular room. She measures the distance with a yardstick and finds it to be 24 yards. The molding is sold in feet. How many feet does she need?

STEP 1 Write the problem.

$$24 \text{ yd} = \underline{\hspace{1cm}} \text{ ft}$$

STEP 2 Write the relationship between yards and feet.

$$1 \text{ yd} = 3 \text{ ft}$$

STEP 3 Multiply the number of yards by 3.

$$3 \times 24 = 72$$

ANSWER: 24 yd = **72 feet**. Toni needs 72 feet of molding for the room.

EXAMPLE 2 Change 96 inches to feet.

STEP 1 Write the problem.

$$96 \text{ in.} = \underline{\hspace{1cm}} \text{ ft}$$

STEP 2 Write the relationship between feet and inches.

$$12 \text{ in.} = 1 \text{ ft}$$

STEP 3 We want to go from 12 inches to 96 inches, and $12 \times 8 = 96$.

$$96 \text{ in.} = 8 \text{ ft}$$

So multiply each side of the equation in Step 2 by 8.

ANSWER: 96 in. = **8 ft**

> In general, when you change from a smaller unit to a larger one, the number of units in your measurement will decrease. When you change from a larger unit to a smaller one, the number of units in your measurement will increase.

Rewrite each measurement so it uses the given unit.

1. 9 ft = _____ in.

2. 2 yd = _____ in.

3. 10 yd = _____ ft

4. 30 ft = _____ in.

5. 4 mi = _____ ft

6. 20 mi = _____ yd

7. 28 yd = _____ ft

8. 10 mi = _____ ft

9. 3 mi = _____ yd

Rewrite each measurement so it uses the given unit.

10. $5\frac{1}{2}$ ft = _____ in.

11. $6\frac{1}{2}$ yd = _____ ft

12. $10\frac{1}{4}$ mi = _____ yd

13. 24 in. = _____ ft

14. 72 in. = _____ yd

15. 33 ft = _____ yd

16. 14 ft = _____ yd

17. 242 ft = _____ yd

18. 6 in. = _____ ft

19. $2\frac{1}{10}$ mi = _____ ft

20. 8.5 ft = _____ in.

21. 7.2 yd = _____ in.

22. 108 in. = _____ yd

23. 72 in. = _____ ft

24. 6 ft = _____ yd

25. 26 in. = _____ ft

26. 30 in. = _____ ft

27. 48 in. = _____ yd

28. $1\frac{1}{4}$ mi = _____ ft

29. $12\frac{1}{3}$ yd = _____ ft

30. $2\frac{1}{6}$ yd = _____ in.

31. 10,560 ft = _____ mi

32. 240 in. = _____ ft

33. 26,400 ft = _____ mi

34. 7 ft = _____ yd

35. 2,640 yd = _____ mi

36. 78 in. = _____ yd

Solve each problem.

37. A plumber has a piece of pipe $2\frac{1}{3}$ yards long. He needs to know the length in feet so he can decide whether it will fit inside his van. What is the length in feet?

38. A paving contractor knows the average number of feet of road his crew can pave per day. He needs to estimate the time it will take to pave a 13-mile section of road. How many feet are in 13 miles?

39. A doctor recommends that her patient walk $1\frac{1}{2}$ miles each day. The patient walks 6 times around a 440-yard track the first day. Did the patient walk $1\frac{1}{2}$ miles?

40. Lee runs a business that makes custom curtains. She needs 240 feet of fabric to fill orders one week. She buys fabric by the yard. How many yards will she have to buy?

41. George is a landscaper. He wants to plant shrubs 1 foot apart in a straight path 66 yards long. He will plant the first and last shrubs half a foot from the ends of the path. How many shrubs will he need?

42. Juan is an installer for an electronics outlet. He is installing a TV in a space that is three and a half feet wide. What is the width of the space in inches?

43. A carpenter needs boards 94 inches long to build a frame for a room. Are 8-foot long boards long enough?

Adding and Subtracting Lengths

Sometimes you have to add or subtract lengths that contain different units.
One way to add and subtract these measurements is to regroup.

EXAMPLE 1 A boiler mechanic used a piece of copper pipe 7 feet 10 inches long and another piece 4 feet 5 inches long for a repair job. How much copper pipe did he use?

STEP 1 Line up the measurements, putting like units under like units.

$$7 \text{ ft } 10 \text{ in.}$$
$$+ \; 4 \text{ ft } \; 5 \text{ in.}$$

STEP 2 Add the inches and add the feet.

$$7 \text{ ft } 10 \text{ in.}$$
$$+ \; 4 \text{ ft } \; 5 \text{ in.}$$
$$11 \text{ ft } 15 \text{ in.}$$

STEP 3 Change the inches to feet and inches.

$$15 \text{ in.} = 1 \text{ ft } 3 \text{ in.}$$

STEP 4 Rewrite the total.

$$11 \text{ ft} + 15 \text{ in.}$$
$$11 \text{ ft} + 1 \text{ ft} + 3 \text{ in.}$$
$$12 \text{ ft } 3 \text{ in.}$$

ANSWER: The mechanic used **12 feet 3 inches** of copper pipe.

Find each sum.

1. 3 yd 2 ft + 6 yd 2 ft

2. 7 ft 10 in. + 2 ft 5 in.

3. 4 yd 21 in. + 3 yd 18 in.

4. 6 ft 11 in. + 2 ft 7 in.

5. 2 yd 8 in. + 3 yd 10 in.

6. 4 yd 28 in. + 2 yd 16 in.

Sometimes you need to add measurements that have two or more units. To add these measurements, the first step is to line up the units that are the same.

EXAMPLE 2 Add 2 ft 8 in. and 2 yd 21 in.

STEP 1 Line up the measurements, putting like units under like units.

$$
\begin{array}{r}
2\,\text{ft}\quad 8\,\text{in.} \\
+\ 2\,\text{yd}\qquad 21\,\text{in.} \\
\hline
\end{array}
$$

STEP 2 Add the like units.

$$
\begin{array}{r}
2\,\text{ft}\quad 8\,\text{in.} \\
+\ 2\,\text{yd}\qquad 21\,\text{in.} \\
\hline
2\,\text{yd}\ 2\,\text{ft}\ 29\,\text{in.}
\end{array}
$$

STEP 3 Change the inches to feet and inches.

29 in. = 24 in. + 5 in.
= 2 ft 5 in.

STEP 4 Rewrite the total.

2 yd + 2 ft + 29 in.
2 yd + 2 ft + 2 ft + 5 in.
2 yd + 4 ft + 5 in.
3 yd + 1 ft + 5 in.

ANSWER: The total is **3 yd 1 ft 5 in.**

EXAMPLE 3 Subtract 1 ft 8 in. from 3 ft 4 in.

STEP 1 Line up the measurements, putting like units under like units.

$$
\begin{array}{r}
3\,\text{ft}\ 4\,\text{in.} \\
-\ 1\,\text{ft}\ 8\,\text{in.} \\
\hline
\end{array}
$$

STEP 2 You have to borrow. The four lines at the right show that 3 ft 4 in. is the same as 2 ft 16 in.

3 ft + 4 in.
2 ft + 1 ft + 4 in.
2 ft + 12 in. + 4 in.
2 ft + 16 in.

STEP 3 Rewrite the problem and subtract.

$$
\begin{array}{r}
2\,\text{ft}\ 16\,\text{in.} \\
-\ 1\,\text{ft}\quad 8\,\text{in.} \\
\hline
1\,\text{ft}\quad 8\,\text{in.}
\end{array}
$$

ANSWER: The difference is **1 ft 8 in.**

EXAMPLE 4   Add 3 yd 28 in. + 6 yd 22 in.

STEP 1   Line up the units. Use "0 ft" for the
number of feet.

$$\begin{array}{r} 3\,\text{yd}\ 0\,\text{ft}\ 28\,\text{in.} \\ +\ 6\,\text{yd}\ 0\,\text{ft}\ 22\,\text{in.} \\ \hline \end{array}$$

STEP 2   Add the yards, feet, and inches.

$$\begin{array}{r} 3\,\text{yd}\ 0\,\text{ft}\ 28\,\text{in.} \\ +\ 6\,\text{yd}\ 0\,\text{ft}\ 22\,\text{in.} \\ \hline 9\,\text{yd}\ 0\,\text{ft}\ 50\,\text{in.} \end{array}$$

STEP 3   Change the inches to yards, feet, and
inches.

50 in. = 36 in. + 12 in. + 2 in.
        = 1 yd 1 ft 2 in.

STEP 4   Rewrite the total.

9 yd + 0 ft + 50 in.
9 yd + 1 yd + 1 ft + 2 in.
10 yd 1 ft 2 in.

ANSWER: The sum is **10 yd 1 ft 2 in.**

Add or subtract.

7. 2 yd 6 ft + 1 ft 12 in.

8. 4 ft 3 in. − 2 ft 7 in.

9. 3 ft 4 in. − 1 ft 11 in.

10. 1 yd 1 ft 11 in. + 3 yd 2 ft 8 in.

11. 4 yd 1 ft − 1 yd 2 ft

12. 6 yd 1 ft 8 in. + 2 yd 2 ft 3 in.

13. 3 ft 8 in. + 2 yd 29 in.

14. 16 yd − 8 yd 2 ft

15. 6 ft 7 in. − 2 ft 9 in.

16. 2 yd 2 ft 14 in. + 3 yd 26 in.

17. 4 ft − 2 ft 5 in.

18. 6 yd 2 ft + 3 ft 10 in.

Solve each problem.

19. An electrician cut a piece of wire 3 feet 7 inches long from a piece that was
7 feet 3 inches long. How long is the piece that she has left?

20. A roofer needs to reach the roof of a two-story building. The heights of the
first and second stories are 12 feet 7 inches and 11 feet 11 inches. What is
the total height?

Multiplying and Dividing Lengths

To multiply or divide with lengths, you use the same two steps but in the opposite order. To multiply a length times a number, first perform the operation and then change the units. To divide a length by a number, first change the units and then perform the operation.

EXAMPLE 1 A display designer wants 4 shelves, each 3 feet 8 inches long, to display samples at a supermarket. What is the total length of board the designer needs?

STEP 1 Decide which math operation to use.

The operation to use is multiplication.

STEP 2 Write the problem.

$$\begin{array}{r} 3 \text{ ft } 8 \text{ in.} \\ \times \qquad 4 \\ \hline \end{array}$$

STEP 3 Multiply the inches and the feet by 4.

$$\begin{array}{r} 3 \text{ ft } 8 \text{ in.} \\ \times \qquad 4 \\ \hline 12 \text{ ft } 32 \text{ in.} \end{array}$$

STEP 4 Change the inches to feet and inches.

$$32 \text{ in.} = 24 \text{ in.} + 8 \text{ in.}$$
$$= 2 \text{ ft } 8 \text{ in.}$$

STEP 5 Rewrite the product.

$12 \text{ ft} + 32 \text{ in.}$
$12 \text{ ft} + 24 \text{ in.} + 8 \text{ in.}$
$14 \text{ ft } 8 \text{ in.}$

ANSWER: The designer needs **14 feet 8 inches** of board in all.

EXAMPLE 2 Divide 9 yards 1 foot by 4.

STEP 1 Write the problem.

$$\frac{9 \text{ yd } 1 \text{ ft}}{4} = ?$$

STEP 2 Change the numerator to feet.

$$9 \text{ yd } 1 \text{ ft} = 27 \text{ ft} + 1 \text{ ft}$$
$$= 28 \text{ ft}$$

STEP 3 Rewrite the problem.

$$\frac{9 \text{ yd } 1 \text{ ft}}{4} = \frac{28 \text{ ft}}{4} = ?$$

STEP 4 Divide. Then change back to yards and feet.

$$\frac{28 \text{ ft}}{4} = 7 \text{ ft}$$
$$= 6 \text{ ft} + 1 \text{ ft}$$
$$= 2 \text{ yd } 1 \text{ ft}$$

ANSWER: The quotient is **2 yards 1 foot**.

Multiply or divide.

1. 6 ft 4 in. × 5 =

2. 6 yd 2 ft ÷ 2 =

3. 10 ft 3 in. ÷ 3 =

4. 12 yd ÷ 2 =

5. 5 yd 1 ft ÷ 4 =

6. 8 in. ÷ 2 =

7. 9 yd 1 ft ÷ 4 =

8. 3 ft 10 in. ÷ 2 =

9. 7 ft 6 in. ÷ 6 =

10. 5 ft 4 in. × 6 =

11. 3 ft 6 in. × 5 =

12. 16 ft 4 in. ÷ 7 =

13. 15 yd ÷ 5 =

14. 10 ft 5 in. ÷ 5 =

15. 6 yd 2 ft × 8 =

16. 4 yd 10 in. × 5 =

17. 6 ft × 5 =

18. 2 ft 4 in. × 8 =

19. 6 ft 4 in. ÷ 4 =

20. 4 ft 9 in. ÷ 3 =

21. 3 ft 11 in. × 0 =

Solve each problem.

22. A cabinet maker wants to make a custom cabinet door by fitting together 4 pieces of wood. She will cut a 10 foot 4 inch board into 4 equal lengths to make the door. How long will each piece be?

23. Mario is a textile designer. He needs 3 yards 8 inches of fabric to cover a chair. If he wants to cover 4 chairs, how much fabric does he need?

24. A machinist needs a steel bar 40 inches long to make a machine part. He needs to make five identical parts. What is the total length of steel he needs? (Answer in feet and inches.)

25. Brian wants to divide a package of insulation into three equal amounts. If the package says $2\frac{1}{2}$ yards, how many feet and inches will each piece be?

Comparing and Ordering Customary and Metric Units of Length

Is a kilometer longer or shorter than a mile? Is an inch longer or shorter than a centimeter? Here are some approximate comparisons that will help you compare the customary units of inches, feet, and miles to the metric units of centimeters, meters, and kilometers.

1 inch	$\approx$	2.54 centimeters	1 centimeter	$\approx \frac{3}{8}$ inch
1 foot	$\approx$	30 centimeters	1 kilometer	$\approx$ 0.6 mile
1 mile	$\approx$	1.6 kilometers		

Misty wants to run a 10-kilometer race, but she is not sure if she can run that far. To find out, she can approximate 10 kilometers with some number of miles.

STEP 1 Write the problem.

$$10 \text{ km} \approx \underline{\hspace{1cm}} \text{ miles}$$

STEP 2 Write the relationship between kilometers and miles.

$$1 \text{ km} \approx 0.6 \text{ mile}$$

STEP 3 We want to go from 1 km to 10 km. So multiply both sides of the equation in Step 2 by 10.

$$10 \text{ km} \approx 6 \text{ miles}$$

ANSWER: If Misty can run about **6 miles**, then she can finish a 10-kilometer race.

Circle the larger measurement.

1. 1 in. or 1 cm

2. 1 m or 1 km

3. 1 cm or 1 yd

4. 1 ft or 1 m

5. 1 cm or 1 ft

6. 1 km or 1 in.

7. 1 mi or 1 km

8. 1 ft or 1 km

9. 1 m or 1 yd

Circle the more appropriate measurement for the length of each object.

10. length of a swimming pool 25 m or 25 in.

11. length of a cell phone 3 in. or 0.5 m

12. length of a beach towel 60 in. or 60 mm

13. length of an umbrella 1 yd or 1 cm

14. width of a business envelope 10 cm or 10 m

15. length of a pencil 15 in. or 15 cm

16. size of a 20-year-old's waist 28 in. or 28 mm

17. length of a queen-size sheet 3 yd or 3 ft

Comparing and Ordering Lengths

To find out which of two lengths is longer or shorter than the other, it is easier if you change the measurements to the same unit.

EXAMPLE 1   Miranda wants to put insulation around a small window. She measured the distance around the window as 130 inches. The insulation comes in packages that contain 3 yards of material. How many packages does she need?

STEP 1 Write the two measurements.

3 yards, 130 inches

STEP 2 Write the relationship between inches and yards.

1 yd = 36 in.

STEP 3 Decide whether to change inches to yards or yards to inches. Then tell how to change from one unit to the other.

Change yards to inches. Multiply the number of yards by 36.

STEP 4 Change yards to inches.

*3 · 36 = 108, so 3 yd = **108** in.*

ANSWER: One package contains 3 yards, or **108 inches**, of insulation. Miranda needs 130 inches of insulation, so she should buy 2 packages.

EXAMPLE 2   Which is shorter, 2.5 meters or 220 centimeters?

STEP 1 Write the two measurements.

2.5 m, 220 cm

STEP 2 Write the relationship between meters and centimeters.

1 m = 100 cm

STEP 3 Decide whether to change meters to centimeters or centimeters to meters. Then tell how to change from one unit to the other.

Change meters to centimeters. Multiply the number of meters by 100.

STEP 4 Change meters to centimeters.

2.5 · 100 = 250, so 2.5 m = 250 cm

STEP 5 Compare and order the measurements.

220 < 250

ANSWER: 220 centimeters is shorter than 2.5 meters.

Circle the smaller measurement.

1. 35 cm or 3 mm

2. 6 ft or 60 in.

3. 65 m or 6 km

4. 2 yd or 7 ft

5. 2,400 m or 3 km

6. 48 in. or 5 ft

7. $6\frac{1}{2}$ m or 570 cm

8. 2 mi or 10,000 ft

9. 3 m or 350 cm

Fill in each blank with <, >, or = to represent "is less than," "is greater than," or "is equal to."

10. 21 ft _____ 13 yd

11. 11 mi _____ 15,000 yd

12. 2 m _____ 200 cm

13. 60 mm _____ 6 cm

14. 1.5 m _____ 7,500 mm

15. 36 in. _____ 3 ft

16. 4 yd _____ 60 in.

17. 2,800 m _____ 3 km

18. 21,000 yd _____ 5 mi

19. 2 in. _____ $\frac{1}{2}$ ft

Write the measurements in order from longest to shortest.

20. 2 in., 2 yd, 2 ft, 2 mi

_____

21. 6 cm, 67 mm, 35 cm, 3 m

_____

22. 26 yd, 18 ft, 4 yd, 3 mi

_____

23. 20 mm, 4 m, 350 cm, 6 cm

_____

Solve each problem.

24. Kai is a carpet installer. He has 5 yards of carpet left on a roll. He needs a 14-foot length to cover a room. Does he have enough?

25. Paul and Ray are each constructing concrete sidewalks at a job site. Paul uses a distance of 1 yard between expansion joints. Ray uses 3 feet between expansion joints. Are they using the same distance? If not, who is using a greater distance?

26. Joan is an insulation worker. She needs 260 inches of insulation to put around a window. She has 6 yards of insulation. Does she have enough?

27. Beth is a salesperson in a store that sells child car seats. A state law requires a certain type of car seat for children between 40 inches and 50 inches tall. Should Beth recommend that type of car seat for a customer who says her child is 4 feet tall?

Focus on Geometry: Measuring Angles

An **angle** is formed when two lines start at the same point. That point is called the **vertex** of the angle, and the lines are called the **sides** of the angle.

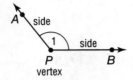

Using the symbol ∠ for angle, you can name an angle three ways:

1. ∠APB   You can name an angle with three letters. The vertex is always the middle letter.

2. ∠P   If a point is the vertex for only one angle, you can name the angle using that single letter.

3. ∠1   You can name an angle by writing a number or letter inside the angle.

Write other names for each angle.

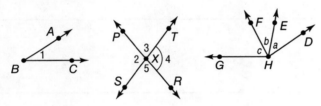

1. ∠1	**3.** ∠3	**5.** ∠5	**7.** ∠b
2. ∠2	**4.** ∠4	**6.** ∠a	**8.** ∠c

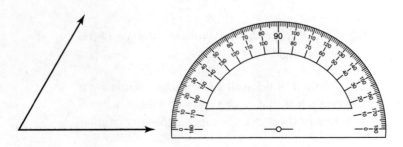

Angles are measured in **degrees**. The tool used to measure an angle is called a **protractor**. A protractor shows measurements along a semicircle from 0° to 180°. It has a dot for the vertex of the angle.

EXAMPLE 1 To use a protractor to measure an angle, follow these steps.

STEP 1 Place the protractor. The vertex of the angle should be under the center of the semicircle. One of the sides of the angle should go through the 0° mark on the protractor.

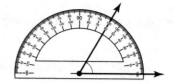

STEP 2 Read the number on the protractor that corresponds to the other side of the angle.

ANSWER: The measurement of the angle is **60°**.

EXAMPLE 2 Using a ruler and a protractor, draw an angle of 70°.

STEP 1 Draw the vertex and one side of the angle.

STEP 2 Place the protractor. The vertex of the angle should be under the center of the semicircle. The side of the angle should go through the 0° mark.

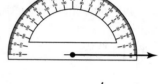

STEP 3 Make a pencil mark where the protractor shows 70°.

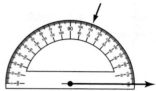

ANSWER: Draw the other side of the angle and label the angle with its measurement, which is 70°.

Use a protractor to measure each angle.

9.

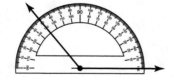

10.

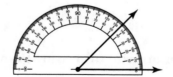

11.

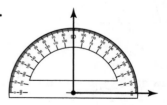

Use a protractor to draw each angle.

12. 120°

13. 115°

14. 25°

15. 90°

16. 180°

17. 65°

Focus on Geometry: Kinds of Angles

There are four different kinds of angles: acute, right, obtuse, and straight.

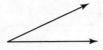

An **acute angle** is an angle between 0° and 90°.

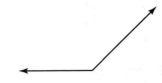

An **obtuse angle** is an angle between 90° and 180°.

A **right angle** is an angle that is exactly 90°. A right angle can be shown with a square corner at the vertex.

A **straight angle** is an angle that is exactly 180°. A straight angle looks like a line.

Label each angle as acute, right, obtuse, or straight. Then use a protractor to measure the angle.

1.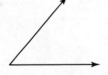

3.

5.

2.

4.

6.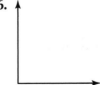

Draw each type of angle. Then use a protractor to measure the angle.

7. an acute angle

9. a straight angle

8. an obtuse angle

10. a right angle

Focus on Geometry: Angles and Parallel Lines

Two lines that are always the same distance apart, and never meet, are called **parallel lines**. The symbol // indicates lines that are parallel, so the expression $\overleftrightarrow{AB}$ // $\overleftrightarrow{CD}$ is read, "line AB is parallel to line CD."

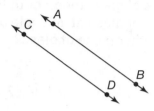

EXAMPLE 1

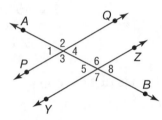

Lines $\overleftrightarrow{PQ}$ and $\overleftrightarrow{YZ}$ are parallel. Which angles are equal to each other?

STEP 1   Use a protractor to measure all the angles.

$\angle 1 = 60°, \angle 2 = 120°$
$\angle 3 = 120°, \angle 4 = 60°$
$\angle 5 = 60°, \angle 6 = 120°$
$\angle 7 = 120°, \angle 8 = 60°$

STEP 2   List the angles that are equal to each other.

ANSWER: $\angle 1 = \angle 4 = \angle 5 = \angle 8, \angle 2 = \angle 3 = \angle 6 = \angle 7$

Some pairs of angles are called **corresponding angles**. Examples are $\angle 1$ and $\angle 5$, $\angle 3$ and $\angle 7$, $\angle 2$ and $\angle 6$, and $\angle 4$ and $\angle 8$. When two parallel lines are cut by a third line, the corresponding angles are always equal.

Some pairs of angles are called **alternate interior angles**. There are two pairs: $\angle 3$ and $\angle 6$, and $\angle 4$ and $\angle 5$. When two parallel lines are cut by a third line, the alternate interior angles are always equal.

The two horizontal lines at the right are parallel. Find the measure of each angle without using a protractor.

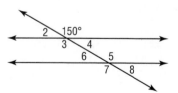

1. $\angle 2 = $ _____

2. $\angle 3 = $ _____

3. $\angle 4 = $ _____

4. $\angle 6 = $ _____

5. $\angle 7 = $ _____

6. $\angle 8 = $ _____

7. $\angle 5 = $ _____

8. $\angle 5 + \angle 6 = $ _____

9. $\angle 3 + \angle 4 = $ _____

Focus on Geometry: Pairs of Angles

When two angles form a right angle or add up to 90°, they are called **complementary angles.** When two angles form a straight angle or add up to 180°, they are called **supplementary angles.**

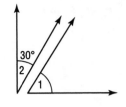

EXAMPLE 1 Josie is a glazier, specializing in installing stained-glass windows. She is fitting two pieces of glass together to form a right angle. If she cuts one piece of glass at an angle of 30°, at what angle should she cut the other piece of glass?

STEP 1 What is the sum of the two angles? $\angle 1 + \angle 2 = 90°$

STEP 2 Substitute 30° for ∠2. $\angle 1 + 30° = 90°$

STEP 3 Subtract 30° from each side. $\angle 1 = 60°$

ANSWER: Josie should cut the other piece of glass at a **60°** angle.

Find the complement of each angle.

1. 75° _____

2. 31° _____

3. 68° _____

4. 47° _____

5. 39° _____

6. 86° _____

Find the complement of each angle.

7.

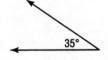

complement: _____

8.

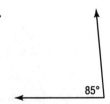

complement: _____

EXAMPLE 2 Find the supplement of an angle whose measure is 50°.

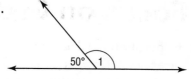

STEP 1 Write the sum of the two angles. $\angle 1 + 50° = 180°$

STEP 2 Subtract 50° from each side. $\angle 1 = 130°$

ANSWER: The measurement of the supplement is **130°**.

Find the supplement of each angle.

9. 80° _____   11. 40° _____   13. 145° _____

10. 120° _____   12. 25° _____   14. 68° _____

Find the supplement of each angle. Describe the supplement as acute, right, obtuse, or straight.

15.

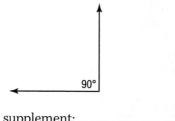

supplement: _____

type: _____

18.

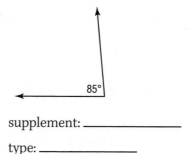

supplement: _____

type: _____

16.

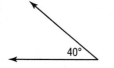

supplement: _____

type: _____

19.

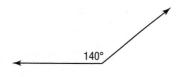

supplement: _____

type: _____

17.

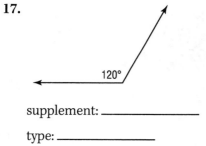

supplement: _____

type: _____

20.

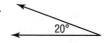

supplement: _____

type: _____

Focus on Geometry: Area and Perimeter

The **perimeter** of a figure is the distance around it.
The **area** of a figure is the size of the region it covers.
Area is measured in square units.

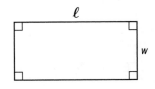

For a rectangle with length ℓ and width w, formulas for
the perimeter P and the area A are:

$$P = 2\ell + 2w \qquad A = \ell \cdot w$$

EXAMPLE 1   Carla's living room is 10 feet by 12 feet. How much wood
trim does she need for the perimeter of the room? How much
carpet does she need for the area of the room?

STEP 1   Write the dimensions of the room. $\ell = 12 \text{ ft}, \qquad w = 10 \text{ ft}$

STEP 2   To find the perimeter, substitute the
values of ℓ and w into the formula
$P = 2\ell + 2w$

$P = 2\ell + 2w$
$P = 2(12) + 2(10)$
$\quad = 24 + 20 = 44 \text{ ft}$

STEP 3   To find the area, substitute the
values of ℓ and w into the formula
$A = \ell \cdot w$.

$A = \ell \cdot w$
$\quad = (12)(10)$
$\quad = 120 \text{ sq ft}$

ANSWER: The perimeter is 44 feet and the area is **120 square feet**.

EXAMPLE 2   Find the perimeter and the area
of this rectangular cloth.

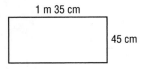

STEP 1   Start by using a single unit
for the length and the width.

Change to centimeters:
$w = 45 \text{ cm}$
$\ell = 1 \text{ m} + 35 \text{ cm}$
$\quad = 100 \text{ cm} + 35 \text{ cm}$
$\quad = 135 \text{ cm}$

STEP 2   To find the perimeter, substitute the
values for ℓ and w into the formula
$P = 2\ell + 2w$.

$P = 2\ell + 2w$
$P = 2(135) + 2(45)$
$\quad = 270 + 90 = 360 \text{ cm}$

STEP 3   To find the area, substitute the
values for ℓ and w into the formula
$A = \ell \cdot w$.

$A = \ell \cdot w$
$\quad = (135)(45)$
$\quad = 6{,}075 \text{ sq cm}$

ANSWER: The perimeter of the cloth is 360 centimeters and
its area is **6,075 square centimeters**.

Find the perimeter and area of each figure.

1. 8 cm
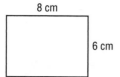
6 cm

P = _____

A = _____

2. 14 ft
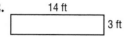
3 ft

P = _____

A = _____

3. 150 mm

3 cm

P = _____

A = _____

4. 7 cm
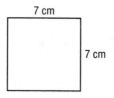
7 cm

P = _____

A = _____

5. 2 ft
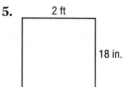
18 in.

P = _____

A = _____

6. 13.5 cm
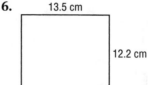
12.2 cm

P = _____

A = _____

7. 3 yd

4 ft

P = _____

A = _____

8. 20 cm
10 mm

P = _____

A = _____

9. 1.2 m
34.5 cm

P = _____

A = _____

Solve each problem.

10. Loretta is an interior designer, planning a design for a room 28 feet long and 20 feet wide. She wants to put crown molding all around the room, so she needs to know its perimeter. What is the perimeter of the room?

11. For the room described in problem 10, Loretta wants to cover the floor with 1-foot by 1-foot tiles. How many tiles does she need?

12. For the room described in problem 10, the ceiling is 8 feet high. If a can of paint covers 300 square feet, how many cans are needed to paint the 4 walls?

13. The distance around a circle is called its **circumference** and the distance across a circle is called its **diameter**. For every circle, the ratio of the circumference to the diameter has the same value, approximately 3.14. That value is represented by the Greek letter π (pi). If r is the radius of a circle, you can use a formula to find the circumference C or the area A:

$$C = 2 \times \pi \times r \qquad A = \pi \times r \times r = \pi r^2$$

Find the circumference and the area of a circle with a radius of 15 centimeters.

Section 2 Cumulative Review

Fill in each blank with the correct measurement.

1. 1 yd = _____ in.

2. _____ ft = 1 mi

3. 1 cm = _____ mm

4. A straight angle contains _____°.

5. 1 ft = _____ in.

6. 1 m = _____ cm

7. A right angle contains _____°.

8. _____ m = 1 km

9. _____ ft = 1 yd

10. The complement of 30° is _____°.

Mark each measurement on the ruler.

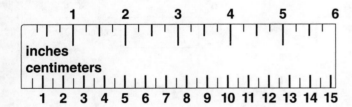

11. $3\frac{1}{4}$ in.

12. 2 cm 9 mm

13. $2\frac{15}{16}$ in.

14. 67 mm

15. 5 cm

16. 4.8 cm

Rewrite each measurement so it uses the given unit.

17. 18 in. = _____ yd

18. 6 m = _____ cm

19. 650 mm = _____ cm

20. 90 in. = _____ ft

21. $3\frac{1}{2}$ ft = _____ in.

22. $8\frac{1}{2}$ yd = _____ ft

23. 108 in. = _____ yd

24. $7\frac{1}{2}$ ft = _____ yd

25. 250 cm = _____ m

26. 3.25 m = _____ cm

Write the measurements from longest to shortest.

27. 38 in., 3 ft, 1 yd + 6 in., 2 yd _____

28. 18 cm, 18 m, 1600 cm, 80 m _____

Use a single unit to find each answer.

29. (6 ft + 8 in.) + (3 ft + 7 in.) = _____

31. $3\frac{1}{2}$ mi ÷ 7 = _____

30. (3 yd + 2 ft) − (1 yd + 17 in.) = _____

32. 5 × 7.2 mi = _____

Label each angle as acute, right, obtuse, or straight. Then use a protractor to measure the angle.

33.

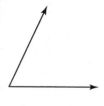

_____

35.

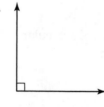

_____

34.

_____

36.

_____

Find the complement of each angle.

37. 62° _____

39. 80° _____

41. 1° _____

38. 17° _____

40. 15° _____

42. 45° _____

Find the supplement of each angle.

43. 115° _____

45. 120° _____

47. 175° _____

44. 40° _____

46. 10° _____

48. 15° _____

Use a protractor to draw the following angles.

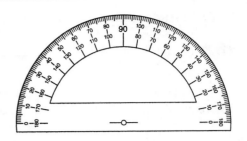

49. 30°

52. 45°

50. 150°

53. 135°

51. 90°

54. 100°

Solve each problem.

55. Find the perimeter and the area of the rectangle.

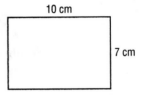

10 cm

7 cm

56. Find the circumference and area of a circular garden with a 10-foot diameter.

57. Antoine is a licensed home improvement contractor, remodeling a house. He needs 20 cedar boards, each $15\frac{1}{2}$ feet long, to use as siding. He needs to know the total length of the boards to calculate how many pounds of nails he will use. What is the total length of the boards?

58. Jane is a seamstress. She needs $3\frac{1}{2}$ yards of fabric to make a special-order dress. How many yards will she need to make eight of the dresses?

Customary Units of Weight

What units do you use when you weigh something? When you use **ounces, pounds, and tons** to weigh an object, you are using customary units of weight. An ounce is a very small customary unit of weight, and a ton is a very large unit. Here are the relationships among these units.

> 1 pound (lb) = 16 ounces (oz)
>
> 1 ton (T) = 2,000 lb

Here are some examples of approximate weights.

1 oz

strawberry
1 ounce

1 lb

can of tomatoes
1 pound

1 T

automobile
1 ton

Name three objects that might weigh the given amount.

1. 1 ounce (1 oz)

_____

2. 1 pound (1 lb)

_____

3. 1 ton (1 T)

_____

For each exercise, circle the heaviest object and underline the lightest object.

4. an orange, a melon, a raspberry

5. a garbage truck, a sport utility vehicle, a motorcycle

6. a glass of juice, a gallon of milk, a liter of juice

7. a paper clip, a ball-point pen, a textbook

8. a 46-inch television, a laptop computer, an MP3 player

Here are some common tools or scales used to measure weights. Some are used to measure very light objects, and others measure very heavy objects. Which of these scales have you used?

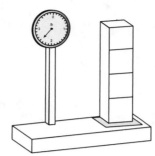

EXAMPLE 1 This scale is in the produce section of a grocery store. Reading the scale is almost like reading a ruler. The scale is labeled in pounds. Each pound is divided into 4 parts, so each tick mark represents 4 ounces. Tai bought some tomatoes and used this scale to weigh them. How much do they weigh?

STEP 1 Decide which units to use for the measurement.

_____ lb and _____ oz

STEP 2 The arrow is between 4 and 5 pounds. Use the smaller number.

4 lb and _____ oz

STEP 3 Count the number of ounces more than 4 pounds.

4 lb and 8 oz

ANSWER: The tomatoes weigh **4 pounds 8 ounces**.

Write the measurement shown on each scale.

9.

10.

11.

_____   _____   _____

Mark the given weight on each scale.

12.

2 lb 15 oz

13.

1 lb 0 oz

14.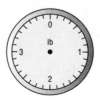

0 lb 7 oz

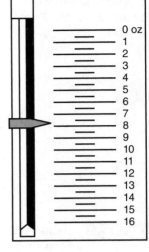

EXAMPLE 2 A nutritionist uses a scale like this to measure small portions of food. The scale is labeled in ounces and half-ounces. How much does this portion of food weigh?

STEP 1 Decide which unit to use for the measurement.

_____ oz

STEP 2 The pointer is between $7\frac{1}{2}$ and 8. Round the weight to the nearest ounce.

8 oz

ANSWER: The portion weighs **8 ounces**.

Write the measurement shown on each scale.

15.

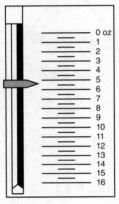

_____

16.

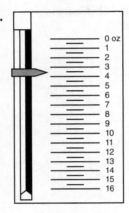

_____

17.

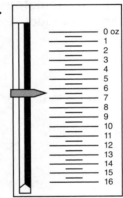

_____

Mark the given weight on each scale.

18.

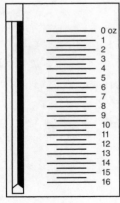

$2\frac{1}{2}$ oz

19.

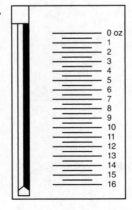

7 oz

20.

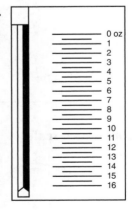

$4\frac{1}{2}$ oz

Metric Units of Weight

Have you ever noticed the weight measurements on a vegetable can? Often there are two measurements on the can, one in ounces and one in grams. A **gram** is a metric unit of weight. Two other metric units of weight are the **milligram** and the **kilogram**.

> 1,000 milligrams (mg) = 1 gram (g)
>
> 1,000 g = 1 kilogram (kg)

In the metric system of measurement, the prefix *milli-* always means $\frac{1}{1000}$ or one-thousandth. The prefix *kilo-* always means 1,000. A very light object is weighed in milligrams, and a heavy object is weighed in kilograms.

a needle weighs
about 300 mg

a peanut weighs
about 1 gram

a telephone book
weighs about 1 kg

Name three things that are sold, measured, or packaged in the given unit.

1. milligrams _____

2. grams _____

3. kilograms _____

Circle the better estimate for the weight of each object.

4. tennis ball   400 mg or 400 g   9. firecracker   30 kg or 30 g

5. bag of apples   3 kg or 3 mg   10. motorcycle   220 g or 220 kg

6. straight pin   200 mg or 200 g   11. stocking hat   800 kg or 800 g

7. 4-year-old girl   15 g or 15 kg   12. feather   100 g or 100 mg

8. sofa   40 mg or 40 kg   13. horse   400 kg or 400 g

Estimating in Ounces, Pounds, and Tons

This lesson shows four different tools for measuring weight. Some tools are used for light objects and some are used for heavy objects.

Scale 1

Scale 2

Scale 1 can be used for portions of food or other light objects. It uses ounces as a unit. Scale 2 often is used to weigh vegetables and fruit. This scale is labeled in pounds. Each pound is divided into 4 parts, so each tick mark represents 4 ounces.

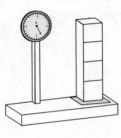

Scale 3

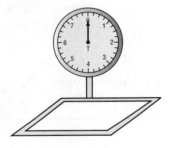

Scale 4

Scale 3 can be used to measure the weight of a package. The scale is labeled in increments of 10 pounds. Each 10-pound increment is divided into 5 parts, so each tick mark represents 2 pounds. Scale 4 is used to measure the weight of cars, trucks, and other heavy objects. The scale is labeled in tons. Each ton is divided into 4 parts, so each tick mark represents 500 pounds.

EXAMPLE **Estimate the weight of an adult's bicycle.**
Then draw an appropriate scale for that weight.
Show your estimated number on the scale.

STEP 1 Think of an appropriate weight.

about 35 lb

STEP 2 Draw and label an appropriate scale for measuring an object with that weight.

ANSWER: The scale shows **35 pounds**.

Estimate the weight of each object. Draw an appropriate scale for that weight. Then show your estimate on the scale.

1. a birthday card to be mailed overseas

2. a crate of canned goods

3. a horse

4. a bag of onions

5. a bag of oranges

6. a stove and a refrigerator

7. a portion of vegetables to be cooked

8. one serving of fish

9. an infant

10. a pickup truck

11. a small package to be mailed

12. a microwave

Changing Among Ounces, Pounds, and Tons

What happens to the weight of an object when you use a different unit to measure it? You know the weight of the object does not change. You will see that if you change from a smaller unit to a larger one, the number of units will decrease. If you change from a larger unit to a smaller one, the number of units will increase.

Here are the relationships among ounces, pounds, and tons.

$$
\begin{aligned}
1 \text{ pound (lb)} &= 16 \text{ ounces (oz)} \\
1 \text{ ton (T)} &= 2{,}000 \text{ lb}
\end{aligned}
$$

EXAMPLE 1 Jackie bought a 5-pound bag of rice. How many ounces of rice did she buy?

STEP 1 Write the problem.

$5 \text{ lb} = \underline{\hspace{1cm}} \text{ oz}$

STEP 2 Write the relationship between ounces and pounds.

$1 \text{ lb} = 16 \text{ oz}$

STEP 3 We want to go from 1 pound to 5 pounds. So multiply both sides of the equation in Step 2 by 5.

$5 \text{ lb} = 80 \text{ oz}$

ANSWER: Jackie has **80 ounces** of rice.

Rewrite each measurement so it uses the given unit.

1. $6 \text{ lb} = \underline{\hspace{1cm}} \text{ oz}$

2. $5 \text{ T} = \underline{\hspace{1cm}} \text{ lb}$

3. $2 \text{ lb} = \underline{\hspace{1cm}} \text{ oz}$

4. $2 \text{ T} = \underline{\hspace{1cm}} \text{ lb}$

5. $3 \text{ T} = \underline{\hspace{1cm}} \text{ lb}$

6. $10 \text{ lb} = \underline{\hspace{1cm}} \text{ oz}$

7. $20 \text{ T} = \underline{\hspace{1cm}} \text{ lb}$

8. $18 \text{ lb} = \underline{\hspace{1cm}} \text{ oz}$

9. $4 \text{ lb} = \underline{\hspace{1cm}} \text{ oz}$

When you change from a smaller unit to a larger one, some of the smaller units may be left over. Here is an example.

EXAMPLE 2 Change 56 ounces to pounds.

STEP 1 Write the problem.

$56 \text{ oz} = \underline{\hspace{1cm}} \text{ lb}$

STEP 2 Write the relationship between ounces and pounds.

$16 \text{ oz} = 1 \text{ lb}$

STEP 3 Divide 56 by 16.

$\dfrac{56}{16} = 3\dfrac{8}{16} = 3\dfrac{1}{2}$

ANSWER: 56 ounces is $3\dfrac{1}{2}$ **pounds**.

Rewrite each measurement so it uses the given unit.

10. 2,500 lb = _____ T

11. 84 oz = _____ lb

12. 22 oz = _____ lb

13. 6,800 lb = _____ T

14. 3,600 lb = _____ T

15. 12 oz = _____ lb

16. 68 oz = _____ lb

17. 8 oz = _____ lb

18. 1,000 lb = _____ T

19. 32 oz = _____ lb

20. 20,000 lb = _____ T

21. 16 oz = _____ lb

22. 8,000 lb = _____ T

23. 80 oz = _____ lb

24. 112 oz = _____ lb

25. 24,000 lb = _____ T

26. 320 oz = _____ lb

27. 40,000 lb = _____ T

28. $1\frac{1}{4}$ lb = _____ oz

29. 9.2 T = _____ lb

30. $2\frac{1}{10}$ T = _____ lb

31. $4\frac{1}{2}$ T = _____ lb

32. 2.3 lb = _____ oz

33. $3\frac{1}{4}$ T = _____ lb

34. $6\frac{1}{2}$ lb = _____ oz

35. $2\frac{3}{4}$ lb = _____ oz

36. 5.5 lb = _____ oz

Use equivalent measurements for ounces, pounds, and tons to solve each problem.

37. Sam, a wholesale produce manager, sold 2.6 tons of produce last week. He wants to calculate the average number of pounds he sold per day. How many pounds is 2.6 tons?

38. Juan manages a restaurant. He bought $12\frac{1}{2}$ pounds of a special seasoning mix at a sale price. He wants to calculate the cost per ounce. How many ounces is $12\frac{1}{2}$ pounds?

39. Fran buys mulch by the ton for her lawn and garden service. She needs 2,800 pounds for a job. How many tons does she need?

40. A can of soup weighs 22 ounces. How many pounds is that?

41. Sue owns a catering service. She wants to make burgers from a 6-pound package of ground meat. How many 4-ounce patties can she make?

42. A municipal park superintendant ordered 22 tons of gravel for a road building project. She needs the weight of the gravel in pounds so she can calculate how many trips it will take to deliver the gravel to the job site. How many pounds is 22 tons?

43. Kim delivered a truckload of cattle. The cattle weighed 36,000 pounds. How many tons did the cattle weigh?

44. Chu bought $1\frac{3}{4}$ pounds of dried fruit. How many ounces did he buy?

45. Tia, a chef, is making a large amount of soup. She multiplies the amounts in the recipe by 10 and finds that she needs 120 ounces of chicken. How many pounds of chicken does she need?

46. George manages a parcel shipping store. The amount he charges is based on weight in pounds and ounces. A customer wants to ship a 90-ounce package. What is that weight in pounds and ounces?

Estimating in Milligrams, Grams, and Kilograms

Here are the approximate weights, in milligrams, grams, and kilograms, of some common objects.

	milligrams	grams	kilograms
a vitamin tablet	1,000 mg	1 g	0.001 kg
a dinner roll	30,000 mg	30 g	0.03 kg
a can of corn		500 g	0.5 kg
a box of laundry soap		1,000 g	1 kg
a large horse		1,000,000 g	1,000 kg

Complete each measurement by writing mg, g, or kg.

1. a hot dog   50 ———

2. a refrigerator   100 ———

3. a small steak   200 ———

4. a brick   1 ———

5. a compact disc   15 ———

6. a tennis ball   175 ———

7. a mini-van   900 ———

8. a cereal serving   30 ———

9. a loaf of bread   0.5 ———

10. a postage stamp   0.1 ———

11. a sandwich   0.25 ———

12. a melon   6 ———

13. a bag of apples   2 ———

14. a stick of gum   1 ———

15. a newspaper   0.4 ———

16. a truck of cement   3,000 ———

Circle the most appropriate unit for measuring the weight of each object.

17. a business envelope   mg   g   kg

18. a jar of salsa   mg   g   kg

19. sodium in a bowl of popcorn mg   g   kg

20. an elephant   mg   g   kg

21. a glass of milk   mg   g   kg

22. a bag of potatoes   mg   g   kg

23. an empty shipping crate   mg   g   kg

Changing Among Milligrams, Grams, and Kilograms

You saw what happened when you used different customary units to weigh an object. The same is true using metric units: if you change from a smaller unit to a larger one, the number of units will decrease. If you change from a larger unit to a smaller one, the number of units will increase.

Here are the relationships among milligrams, grams, and kilograms.

1 mg	=	0.001 g
1,000 mg	=	1 g
1 g	=	0.001 kg
1,000 g	=	1 kg

EXAMPLE 1 A cleaning crew needs to use 2 kilograms of powdered disinfectant to prepare the surgery suite of a hospital. How many grams of disinfectant are they going to use?

STEP 1 Write the problem.

$2 \text{ kg} = \underline{\quad} g$

STEP 2 Write the relationship between kilograms and grams.

$1 \text{ kg} = 1,000 \text{ g}$

STEP 3 We want to go from 1 kilogram to 2 kilograms. So multiply both sides of the equation in Step 2 by 2.

$2 \text{ kg} = \textbf{2,000 g}$

ANSWER: They will use **2,000 grams** of disinfectant.

Rewrite each measurement so it uses the given unit.

1. $6 \text{ kg} = \underline{\quad} g$

2. $27 \text{ kg} = \underline{\quad} g$

3. $18 \text{ g} = \underline{\quad} mg$

4. $17 \text{ kg} = \underline{\quad} g$

5. $245 \text{ g} = \underline{\quad} mg$

6. $3 \text{ g} = \underline{\quad} mg$

Sometimes a measurement contains a mixed number. Here is an example.

EXAMPLE 2 Change $2\frac{1}{2}$ kilograms to grams.

STEP 1 Write the problem.

$2\frac{1}{2} \text{ kg} = \underline{\quad} g$

STEP 2 Write the relationship between kg and g.

$1 \text{ kg} = 1,000 \text{ g}$

STEP 3 We want to go from 1 kilogram to $2\frac{1}{2}$ kilograms. So multiply each side of the equation in Step 2 by $2\frac{1}{2}$.

$2\frac{1}{2} \text{ kg} = \textbf{2,500 g}$

ANSWER: $2\frac{1}{2} \text{ kg} = \textbf{2,500 g}$

Rewrite each measurement so it uses the given unit.

7. 3.6 g = _____ mg

8. 8.6 kg = _____ g

9. $6\frac{3}{10}$ kg = _____ g

10. $4\frac{1}{4}$ kg = _____ g

11. 12.6 g = _____ mg

12. 2.45 kg = _____ g

13. $6\frac{2}{5}$ g = _____ mg

14. 7.25 g = _____ mg

EXAMPLE 3 Ethel, a community health educator, is teaching a class about healthy eating habits. She tells the class that many pizzas have 2,500 milligrams of sodium or more. She wants to show that amount of sodium on a scale that weighs items in grams. How many grams of sodium should Ethel put on the scale?

STEP 1 Write the problem.

STEP 2 Write the relationship between g and mg.

STEP 3 We want to go from 1,000 milligrams to 2,500 milligrams. So multiply both sides of the equation in Step 2 by 2.5.

2,500 mg = _____ g

1,000 mg = 1 g

2,500 mg = 2.5 g

ANSWER: Ethel should put **2.5 grams** of sodium on the scale.

Rewrite each measurement so it uses the given unit.

15. 20,000 g = _____ kg

16. 6,000 mg = _____ g

17. 7,000 mg = _____ g

18. 8,000 g = _____ kg

19. 2,000 g = _____ kg

20. 26,000 mg = _____ g

21. 2,100 mg = _____ g

22. 4,725 mg = _____ g

23. 2,750 g = _____ kg

24. 4,200 g = _____ kg

25. 325 mg = _____ g

26. 3,600 g = _____ kg

Use the equivalent measurements for milligrams, grams, and kilograms to solve each problem.

27. A frozen dinner has $3\frac{1}{2}$ grams of fat. How many milligrams of fat does it contain?

28. George, a purchaser for a hotel kitchen, bought a 9.1-kilogram bag of cornmeal. He wants to tell the cook staff the weight of the cornmeal in grams to help them plan their menus. What is the weight in grams?

29. Sheila weighed a portion of vegetables. It weighed 90 grams. How many milligrams is the portion of vegetables?

30. Mateo, a hospital dietician, wants his patient to consume fewer than 1.5 grams of sodium daily. Some food packages are labeled in milligrams, so he needs to tell the patient the amount in milligrams. How many milligrams is 1.5 grams?

31. Julia owns a health food store. She has 4 kilograms of organic rice. How many 500-gram packages of rice can she make?

32. Kim bought 1.75 kilograms of candy. How many grams is that?

33. Mai, a certified athletic trainer, learned that some experts recommend as much as 4,700 milligrams of potassium per day in an athlete's diet. She wants to show that amount to a client in grams. How many grams is 4,700 milligrams?

34. Hector weighed a piece of chicken at 100 grams. How many kilograms is that?

35. Koto is teaching a unit on nutrition to her high school science students. She says an average-size man can eat 56,000 milligrams of fat per week as part of a healthy diet. She wants to show 56,000 milligrams on a scale that weighs in grams. How many grams is 56,000 milligrams?

36. Joi is a shipping clerk. She weighed a package as 1,750 grams. She needs to label the package weight in kilograms. How many kilograms does the package weigh?

37. A packing company put glasses in very small boxes. Each box weighed 3,500 grams. How many kilograms did each box weigh?

38. Sara is a health and wellness director for a large corporation. During a seminar, she recommends that adults eat about 32,500 milligrams of fiber daily. A person in the audience asks what that amount is in grams. What should be Sara's answer?

Adding and Subtracting Weights

When you work with measurements, sometimes you need to add or subtract the measurements. Adding or subtracting measurements can involve regrouping and borrowing.

EXAMPLE 1 Leslie is a medical assistant in a pediatrician's office. She tells a woman that her baby has gained 2 pounds 7 ounces since birth. If the baby weighed 6 pounds 12 ounces at birth, how much does the baby weigh now?

STEP 1 Line up the measurements, putting like units under like units.

$$\begin{array}{r} 6 \text{ lb} \quad 12 \text{ oz} \\ + 2 \text{ lb} \quad 7 \text{ oz} \\ \hline \end{array}$$

STEP 2 Add the ounces and add the pounds.

$$\begin{array}{r} 6 \text{ lb} \quad 12 \text{ oz} \\ + 2 \text{ lb} \quad 7 \text{ oz} \\ \hline 8 \text{ lb} \quad 19 \text{ oz} \end{array}$$

STEP 3 Change 19 ounces to pounds and ounces.

$$19 \text{ oz} = 16 \text{ oz} + 3 \text{ oz}$$
$$= 1 \text{ lb } 3 \text{ oz}$$

STEP 4 Rewrite the total.

$$8 \text{ lb} + 19 \text{ oz} = 8 \text{ lb} + 1 \text{ lb} + 3 \text{ oz}$$
$$= 9 \text{ lb } 3 \text{ oz}$$

ANSWER: The baby now weighs **9 pounds 3 ounces**.

EXAMPLE 2 Subtract 9 pounds 17 ounces from 13 pounds 4 ounces.

STEP 1 Line up the measurements, putting like units under like units.

$$\begin{array}{r} 13 \text{ lb} \quad 4 \text{ oz} \\ - 9 \text{ lb} \quad 17 \text{ oz} \\ \hline \end{array}$$

STEP 2 Think of 13 lb as 12 lb 16 oz. Then rewrite 13 lb 4 oz as 12 lb 20 oz.

$$\begin{array}{r} {}^{12} \quad {}^{20} \\ \cancel{13} \text{ lb} \quad \cancel{4} \text{ oz} \\ - 9 \text{ lb} \quad 17 \text{ oz} \\ \hline \end{array}$$

STEP 3 Subtract the pounds and subtract the ounces.

$$\begin{array}{r} 12 \text{ lb} \quad 20 \text{ oz} \\ - 9 \text{ lb} \quad 17 \text{ oz} \\ \hline 3 \text{ lb} \quad 3 \text{ oz} \end{array}$$

ANSWER: The difference is **3 pounds 3 ounces**.

Add or subtract.

1. 6 lb 13 oz + 4 lb 7 oz

2. 4 lb 2 oz − 2 lb 12 oz

3. 18 lb 12 oz + 6 lb 9 oz

4. 7 lb 12 oz − 3 lb 4 oz

5. 12 lb 7 oz − 8 lb 2 oz

6. 2 lb 7 oz + 4 lb 3 oz

7. 3 lb 6 oz + 7 lb 15 oz

8. 2 lb 4 oz − 1 lb 10 oz

9. 16 lb 2 oz − 10 lb 4 oz

10. 12 lb 2 oz + 4 lb 3 oz

11. 8 lb − 3 lb 7 oz

12. 2 lb 14 oz + 1 lb 7 oz

To add or subtract metric measurements, make sure the measurements use the same units.

EXAMPLE 3 Find the sum and the difference of 5.32 kilograms and 150 grams.

STEP 1 Rewrite the measurements using the same unit.

$$5.32 \text{ kg} = 5,320 \text{ g}$$

STEP 2 Line up the two measurements. Then add or subtract.

Sum	Difference
5,320	5,320
+ 150	− 150
5,470	5,170

ANSWER: The sum is **5,470 grams** (or 5.47 kilograms).
The difference is **5,170 grams** (or 5.17 kilograms).

Find each sum or difference.

13. 36 g + 2 kg

14. 15 g − 400 mg

15. 12 kg − 750 g

16. 12 mg + 12 g

17. 1 kg + 1 g + 1 mg

18. 1 kg + 1,000 g + 1,000,000 mg

Solve each problem.

19. James works in a kennel. He has a container with 18 pounds 12 ounces of dog food. He uses 6 pounds 4 ounces of the dog food for a morning feeding. How much is left?

20. Carley sells products from home. She is having a party for customers and plans to make a dish that calls for 4 pounds of chicken. She bought a package that weighs 2 pounds 7 ounces and another that weighs 1 pound 12 ounces. Did she buy enough?

Multiplying and Dividing Weights

Sometimes you need to multiply or divide weights. If a large carton contains many of the same item, you can use division to find the weight of each item. If you are shipping many packages that have the same weight, you can use multiplication to find the total weight.

EXAMPLE 1 Rachel is a summer camp supervisor. One of her responsibilities is to plan meals and buy food. She bought a "family pack" of lean ground beef that contains 5 packages. Each package weighs 4 pounds 9 ounces. What is the total weight of the ground beef?

STEP 1 Decide which math operation to use.

The operation to use is multiplication.

STEP 2 Write the problem.

$$4 \text{ lb } 9 \text{ oz}$$
$$\times \quad 5$$

STEP 3 Multiply the ounces and the pounds by 5.

$$4 \text{ lb } 9 \text{ oz}$$
$$\times \quad 5$$
$$\overline{20 \text{ lb } 45 \text{ oz}}$$

STEP 4 45 ounces is the same as 32 ounces plus 13 ounces. So rewrite 20 pounds 45 ounces.

$$20 \text{ lb } + 45 \text{ oz}$$
$$= 20 \text{ lb } + 2 \text{ lb } + 13 \text{ oz}$$
$$= 22 \text{ lb } + 13 \text{ oz}$$

ANSWER: The total weight is **22 pounds 13 ounces**.

EXAMPLE 2 Rachel's campers have a favorite dish that uses 3.5 pounds of ground beef. How can Rachel use her calculator to find out how many times the cook can make that dish, using 22 pounds 13 ounces of ground beef?

STEP 1 Decide which math operation to use.

The operation to use is division.

STEP 2 Use a calculator to write 22 pounds 13 ounces as a decimal.

$$22 \text{ lb } 13 \text{ oz} = 22 + \frac{13}{16}$$
$$= 22 + 0.8125$$
$$= 22.8125$$

STEP 3 Use a calculator to divide 22.8125 by 3.5.

$$22.8125 \div 3.5 = 6.5179$$

ANSWER: The cook can make the dish **6 times**. (There will be enough left over for another half batch of the dish.)

Multiply or divide.

1. 2 lb 5 oz × 3

2. 4 lb 5 oz ÷ 3

3. 1 lb 3 oz × 8

4. 6 lb 2 oz ÷ 2

5. 12 lb 3 oz ÷ 5

6. 7 lb 10 oz × 4

7. 6 oz × 4

8. 3 lb 8 oz ÷ 2

9. 12 lb 5 oz × 6

10. 22 lb 2 oz ÷ 6

11. 24 lb 7 oz × 4

12. 38 lb 1 oz ÷ 7

13. 10 lb 1 oz × 3

14. 6 lb 12 oz ÷ 6

Solve each problem.

15. Alicia employs home caregivers and sends them to seniors' homes to cook, clean, wash clothes, and do other tasks. She has a box of detergent that weighs 40 pounds 10 ounces. She wants to put the detergent into 5 equal packages for her 5 employees. How much should each package weigh?

16. Elena operates a cleaning service. She wants to mail the same gift to each of 7 new clients. Each gift weighs 2 pounds 3 ounces, and the mailing cost depends on the total weight. What is the total weight of all 7 gifts?

17. Henri is a caterer. He has 7 pounds 12 ounces of ham to feed 31 people. How many ounces can he serve each person?

18. A cook needs to know the weight of her ingredients. She used 5 cans of beans for chili. Each can holds 1 pound 2 ounces of beans. What was the total weight of the beans?

Comparing and Ordering Weights

Sometimes you want to know which of two objects is lighter or heavier. It is easier to compare weights if you change the measurements to the same unit. When you compare measurements, you can use the math symbols $<$ or $>$ to represent "is less than" or "is greater than."

EXAMPLE 1 Maria is a cook's helper in an assisted living facility. The cook asks her to use a recipe that calls for 24 ounces of rice. She has a 2-pound package of rice. Does she have enough rice for the recipe?

STEP 1   Write the two measurements. $2\ lb;\ 24\ oz$

STEP 2   Write the relationship between pounds $1\ lb = 16\ oz$
         and ounces.

STEP 3   Change one of the measurements. $2\ lb = (2)(16) = 32\ oz$

STEP 4   Compare the measurements. $32\ oz > 24\ oz$

ANSWER: Maria has **32 ounces** of rice, which is enough for the recipe.

EXAMPLE 2   Which is smaller, 3,500 milligrams or 3 grams?

STEP 1   Write the two measurements. $3,500\ mg;\ 3\ g$

STEP 2   Write the relationship between $1,000\ mg = 1\ g$
         milligrams and grams.

STEP 3   Change one of the measurements. $3,500\ mg = \frac{3,500}{1,000}\ g = 3.5\ g$

STEP 4   Compare the measurements. $3\ g < 3.5\ g$

ANSWER: **3 grams** is smaller than 3,500 milligrams.

Circle the smaller measurement.

1. 3 lb or 50 oz

2. 3.5 kg or 4,200 g

3. $2\frac{1}{2}$ lb or 42 oz

4. 3 T or 4,800 lb

5. 128 oz or $7\frac{1}{2}$ lb

6. 2,700 mg or 2.5 g

7. 18 oz or 2 lb

8. 4 kg or 360 g

Fill in each blank with >, <, or =.

9. 6 oz _____ 1 lb

10. 2 T _____ 800 lb

11. 420 mg _____ 3 g

12. 3 kg _____ 3,000 g

13. $2\frac{1}{2}$ T _____ 7,000 lb

14. 124 oz _____ 7 lb

15. 3 kg _____ 300 mg

16. 6 lb _____ 96 oz

Write the measurements from lightest to heaviest.

17. 14 oz, 1.5 lb, 7 oz, 2 lb

_____

18. 18 lb, 3 T, 270 lb, 1.6 T

_____

19. 16 mg, 6 kg, 6 g, 482 g

_____

20. 800 g, 240 mg, 1 kg, 45 g

_____

Solve each problem.

21. Kayla supplies natural food products to retail shops. The Coffee Café and The Natural Home Bake Shop each need granola. She has two packages; one is labeled 42 ounces and the other is labeled 3 pounds. Kayla wants to send the larger package to The Coffee Café. Which package should she send to the Coffee Café?

22. Tia is a delivery room nurse. She tells a new father that his son weighs $7\frac{1}{2}$ pounds. The father asks if that is 7 pounds 8 ounces. How should Tia answer?

23. Sam shops and runs errands for disabled customers. He needs to buy a gift and mail it for a customer. He is comparing a box of candy labeled 22 ounces and a box of candy labeled 1 pound 4 ounces. For the mailing cost, he wants to know which box weighs more. Which is greater, 22 ounces or 1 pound 4 ounces?

24. A school cook is comparing the fat content in two soup recipes. One is 9 grams per serving and the other is 400 milligrams per serving. Which amount is smaller?

Comparing Customary and Metric Units of Weight

Sometimes the weight of an object is given in both customary units and in metric units. For example, a candy bar wrapper might list the weight as:

net wt. 1.6 oz (45 g)

Ounces, pounds, and tons are the basic units of weight in the customary measurement system. Milligrams, grams, and kilograms are the basic units of weight in the metric system.

Here are some equivalent measurements for customary and metric units of weight.

16 oz = 1 lb	1,000 mg = 1 g	1 kg ≈ 2.2 lb
2,000 lb = 1 T	1 mg = 0.001 g	1 oz ≈ 28 g
	1,000 g = 1 kg	1 T ≈ 909 kg

EXAMPLE 1   Jim teaches English language learners in an evening school. He wants to teach his students how to compare an ounce and a gram. Which is a greater weight?

STEP 1   Write the two units.

$1 \ oz; 1 \ g$

STEP 2   Write the relationship between the two units.

$1 \ oz \approx 28 \ g$

STEP 3   Decide which is larger.

It takes 28 grams to make 1 ounce, so 1 gram is much smaller than 1 ounce.

ANSWER: An ounce is larger than a gram.
(An ounce is about 28 times as much as a gram.)

Circle the larger measurement.

1. 1 gram   or   1 lb

2. 1 ton   or   1 lb

3. 1 kg   or   1 lb

4. 1 g   or   1 oz

5. 1 oz   or   1 lb

6. 1 ton   or   1 g

7. 1 oz   or   1 kg

8. 1 mg   or   1 g

9. 1 kg   or   1 ton

10. 1 mg   or   1 oz

Name three objects whose weight falls between the given amounts.

11. 1 gram and 50 grams

_____

12. 1 ton and 5 tons

_____

13. 1 pound and 5 pounds

_____

14. 1 kilogram and 5 kilograms

_____

15. Where might you see weight of an object listed using both customary units and metric units?

_____

Circle the more appropriate measurement for the weight of each object.

16. a strawberry
   6 oz or 6 g

17. a bowling ball
   3 oz or 3 kg

18. a can of green beans
   1 g or 1 lb

19. a child
   24 kg or 24 oz

20. a cow
   1 kg or 1 ton

21. a paper clip
   400 mg or 400 g

Use the information on customary and metric units of weight to answer each question.

22. An object weighs 14 grams. What fraction of an ounce does it weigh?

23. A book weighs 1 pound. Does it weigh more or less than half a kilogram?

24. Is 1,000 ounces more or less than 1 kilogram?

25. Is 200 grams more or less than 1 pound?

26. About how many pounds does it take to balance
1 kilogram + 1 gram + 1 milligram?

27. About how many kilograms does it take to balance
1 ton + 1 pound + 1 ounce?

Focus on Thermometers: Fahrenheit and Celsius Scales

You probably have seen the two different thermometers that are used to measure temperature. One measures the temperature in degrees Fahrenheit (°F). The other measures the temperature in degrees Celsius (°C).

This thermometer shows both the Fahrenheit and Celsius scales. Some of the most common measurements are shown.

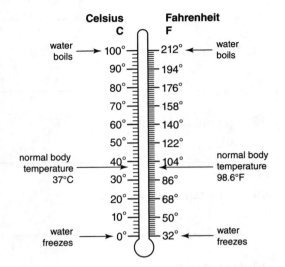

EXAMPLE 1   **At what temperature does water freeze?**

STEP 1   Find the two labels in the diagram that say "water freezes."

STEP 2   Look at the labels on the top of each scale for the words "Celsius" and "Fahrenheit."

ANSWER: **When water freezes, the reading on a Celsius thermometer shows 0°C. The reading on a Fahrenheit thermometer shows 32°F.**

Use the Celsius and Fahrenheit scales on the thermometer to answer each question.

1. What is normal body temperature?

    _____ °C   _____ °F

2. At what temperature does water boil?

    _____ °C   _____ °F

3. How many degrees are there between the freezing and boiling points of water?

    _____ °C   _____ °F

Use the thermometer above to fill in the chart.

	°F	°C
4.		30°C
5.		80°C
6.	104°F	
7.	194°F	
8.		50°C

Circle the warmer temperature.

9. 28°C or 28°F

10. 90°C or 90°F

11. 72°F or 24°C

12. 60°F or 48°C

Circle the more reasonable temperature for each.

13. ice cream
30°C or 30°F

14. swimming pool water
70°C or 70°F

15. comfortable
sleeping room
68°C or 68°F

16. a cup of hot tea
80°C or 80°F

17. sleet
28°F or 28°C

18. a cool autumn day
8°F or 8°C

19. normal body
temperature
98.6°F or 98.6°C

20. a cold winter's day
12°C or 12°F

Circle the colder temperature.

21. 6°C or 6°F

22. 0°C or 0°F

23. 29°C or 29°F

24. 80°C or 150°F

25. 75°C or 104°F

26. 20°C or 80°F

In the diagram below, the temperature 14°C is marked on the thermometer. Mark the given temperatures on the thermometer.

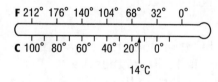

27. 70°F   **28.** 180°F   **29.** 65°C   **30.** 42°C

Focus on Thermometers: Temperatures

Sometimes a meteorologist on the radio or television reports the current temperature and tells you there will be a particular temperature change. You can use addition or subtraction to find the expected, changed temperature.

 A meteorologist, showing this thermometer, says that the morning temperature is 40°F and that the temperature will fall 8 degrees during the day. What is the temperature expected at the end of the day?

STEP 1 Write the morning temperature. $40°F$

STEP 2 "Fall" means "become lower." So "fall 8 degrees" means to subtract 8 from 40. $40 - 8$

STEP 3 Subtract. $40 - 8 = 32$

ANSWER: The temperature at the end of the day is expected to be **32°F**.

Find the result for each temperature change.

1. Current temperature: 72°F
   Change: 12° drop

2. Current temperature: 38°C
   Change: 10° increase

3. Current temperature: 36°F
   Change: 17° drop

4. Current temperature: 25°C
   Change: 6° drop

5. Current temperature: 92°F
   Change: 6° decrease

6. Current temperature: 14°C
   Change: 3° rise

On a cold day or night, the temperature may be below zero.

 A meteorologist, showing this thermometer, says that the evening temperature is 12°F and that the temperature will fall 17 degrees during the night. What is the coldest temperature expected during the night?

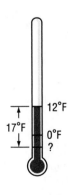

STEP 1 Write the evening temperature. $12°F$

STEP 2 "Fall" means "become lower." So "fall 17 degrees" means to subtract 17 from 12. $12 - 17$

STEP 3 Subtract. $12 - 17 = -5$

ANSWER: The coldest temperature expected during the night is **−5°F**.

Find the result for each temperature change.

7. Current temperature: 5°C
   Change: 8° drop

11. Current temperature: 0°F
    Change: 3° increase

8. Current temperature: −14°C
   Change: 7° increase

12. Current temperature: 21°F
    Change: 36° drop

9. Current temperature: 0°F
   Change: 6° drop

13. Current temperature: 18°F
    Change: 21° fall

10. Current temperature: 21°C
   Change: 6° drop

14. Current temperature: 3°C
    Change: 7° drop

Circle the colder temperature. You can look at a thermometer to help you decide.

15. −8°F or 8°F

18. 0°F or 6°F

16. 15°C or −16°C

19. 4°C or −7°C

17. 100°F or 32°C
(Be careful comparing
°F and °C.)

20. 0°C or 30°F
(Be careful comparing
°F and °C.)

Mark the given temperatures on the Fahrenheit thermometer.

21. 36°F

22. −12°F

23. 48°F

24. 0°F

25. 45°F

26. −5°F

27. −17°F

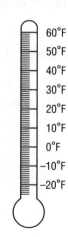

Focus on Thermometers: Changing Between °F and °C

There are two formulas for changing between Fahrenheit and Celsius temperatures.

Change from C to F:	Change from F to C:
$F = \frac{9}{5}C + 32$	$C = \frac{5}{9}(F - 32)$

EXAMPLE 1 The outdoor thermometer at the bank says the temperature is 25°C. Manuel wants to know what the Fahrenheit temperature is.

STEP 1 Write the formula for changing Celsius to Fahrenheit.

$$F = \frac{9}{5}C + 32$$

STEP 2 Substitute the given value of C into the formula.

$$F = \frac{9}{5} \cdot 25 + 32$$

STEP 3 Find the value of F.

$$= \frac{9}{\underset{1}{\cancel{5}}} \cdot \overset{5}{\cancel{25}} + 32$$
$$= 45 + 32$$
$$= 77$$

ANSWER: 25°C = **77°F**

Change each Celsius temperature to a Fahrenheit temperature.

1. 30°C = _____ °F

2. 40°C = _____ °F

3. 75°C = _____ °F

4. 10°C = _____ °F

5. 45°C = _____ °F

6. 90°C = _____ °F

EXAMPLE 2 The temperature is 50°F. What is the temperature on a Celsius thermometer?

STEP 1 Write the formula for changing Fahrenheit to Celsius.

$$C = \frac{5}{9}(F - 32)$$

STEP 2 Substitute the given value of F into the formula.

$$C = \frac{5}{9}(50 - 32)$$

STEP 3 Find the value of C.

$$C = \frac{5}{9} \cdot 18$$
$$C = \frac{5}{\underset{1}{\cancel{9}}} \cdot \overset{2}{\cancel{18}}$$
$$= 10$$

ANSWER: 50°F = **10°C**

Change each Fahrenheit temperature to a Celsius temperature.

7. 104°F = _____ °C

8. 158°F = _____ °C

9. 68°F = _____ °C

10. 32°F = _____ °C

11. 72°F = _____ °C

12. 86°F = _____ °C

13. −4°F = _____ °C

14. 59°F = _____ °C

15. 212°F = _____ °C

16. −40°F = _____ °C

Use $F = \frac{9}{5}C + 32$ or $C = \frac{5}{9}(F - 32)$ to solve each problem.

17. Water boils at 100 degrees Celsius. What Fahrenheit temperature is that?

18. Water freezes at 0 degrees Celsius. What Fahrenheit temperature is that?

19. By putting salt on a sidewalk, snow will melt at or above 23 degrees Fahrenheit. What Celsius temperature is that?

20. A maintenance supervisor for an apartment building installs an air conditioner that has a thermostat marked in degrees Celsius. He wants to set the thermostat at the Celsius equivalent of 77 degrees Fahrenheit. What Celsius setting should he use?

21. A heater's thermostat is set at 68 degrees Fahrenheit. What Celsius temperature is that?

22. Kevin is a registered nurse. A patient's chart says to notify a doctor immediately if the patient's temperature reaches 104 degrees Fahrenheit. Kevin has a thermometer marked in degrees Celsius. At what Celsius temperature must Kevin notify a doctor?

23. Emily is a hotel worker. She normally works in the U.S. but is now working in Mexico. In the U.S. she sets hotel room refrigerator temperatures at 41 degrees Fahrenheit. What Celsius temperature should she use while working in Mexico?

24. A case of chocolate from Europe has a warning on the side: "Do not store above 30 degrees Celsius." At what Fahrenheit temperature should the chocolate be stored?

Section 3 Cumulative Review

Rewrite each measurement so it uses the given unit.

1. 1 lb = _____ oz

2. 1 kg ≈ _____ lb

3. 1 g = _____ mg

4. 1 T = _____ lb

5. 1 oz ≈ _____ g

6. 66 oz = _____ lb

7. 4,500 lb = _____ T

8. 3,500 mg = _____ g

9. $2\frac{1}{2}$ lb ≈ _____ g

10. 4.3 kg = _____ g

Circle the larger measurement.

11. 2 lb or 2 kg

12. 32°F or 32°C

13. 600 g or 600 mg

14. 0°C or 10°F

15. $2\frac{1}{2}$ tons or $2\frac{1}{2}$ kilograms

16. 15 oz or 15 g

Circle the better estimate for each object.

17. temperature of bath water   80°C or 80°F

18. weight of a can of soup   16 oz or 16 g

19. temperature of a popsicle   26°C or 26°F

20. weight of a phone book   1 g or 1 kg

21. weight of an elephant   2.2 kg or 2.2 tons

Mark the given weight on each scale.

22.

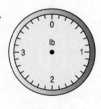

2 lb 3 oz

23.

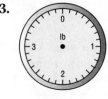

$1\frac{1}{2}$ lb

24.

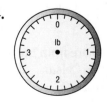

3 lb 13 oz

Write the measurements from lightest to heaviest.

25. 26 g, 26 lb, 26 oz, 26 kg _____

26. 2 T, 2 oz, 2 lb _____

27. 1 kg, 1 mg, 1 g _____

Solve each problem.

28. Jason owns a candy and gift shop. He has to divide 18 pounds of candy into 8 equal portions to make gift boxes. What is the weight of each portion?

29. Jackie works at a delicatessen. She used 2 pounds 7 ounces of ham and 3 pounds 4 ounces of turkey to fill a lunch order. How much meat did she use?

30. The total weight of three packages is 4 pounds 5 ounces. If one package weighs 2 pounds 7 ounces, what is the total of the other two?

31. Chu made 7 bags of his specialty cereal to sell in his health food store. Each bag had 12 ounces of cereal. How much cereal did he use in all? Give your answer in pounds and ounces.

32. What is the normal body temperature in degrees Fahrenheit?

33. Ingrid is a tour guide in New York. She tells her group the temperature is 59°F. A tourist from Europe asks for the Celsius temperature. What should be Ingrid's answer?

34. During a bad cold, the Smith baby lost 1 pound 5 ounces. After the cold she weighed 13 pounds 3 ounces. How much did she weigh before she caught the cold?

35. Robin works at a parcel shipping store. A customer wants to mail 2 packages. One weighs 2 pounds 12 ounces, and the other weighs 1 pound 14 ounces. Robin bases the shipping cost on the combined weight. What is the combined weight of the packages?

36. Change 35°C to °F.

37. What is the freezing point of water?

38. What is the boiling point of water?

39. Which is warmer, 20°F or 20°C?

Mark the given temperatures on the Fahrenheit thermometer.

40. 103°F

41. 97°F

42. 102.5°F

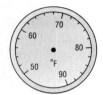

Mark the given temperatures on the Celsius thermometer.

43. 24°C   **45.** 7°C

44. 12°C   **46.** −5°C

At the right is a thermostat for a furnace.

47. At what temperature do you set your thermostat in the winter for a comfortable daytime temperature? Mark the thermostat.

_____

48. At what temperature do you set your thermostat in the winter for a comfortable sleeping temperature? Mark the thermostat.

_____

Mark the given temperatures on the thermostat.

49. 72°F   **51.** 78°F

50. 54°F   **52.** 69°F

Below is an outdoor Celsius thermometer. It measures how warm the air is. Mark the given temperatures on the thermometer.

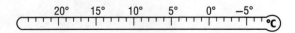

53. 16°C   **54.** 2°C   **55.** 21°C   **56.** −5°C

Answer each question.

57. What Celsius reading is a comfortable outside temperature? _____

58. At what Celsius temperature might it be snowing? _____

List the names of the units.

59. List four basic customary units for measuring length.

_____

60. List three basic customary units for measuring weight.

_____

61. List three basic metric units for measuring length.

_____

62. List three basic metric units for measuring weight.

_____

Write the measurement shown.

63.

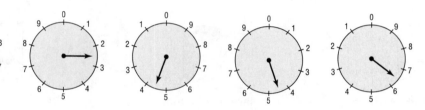

_____

64.

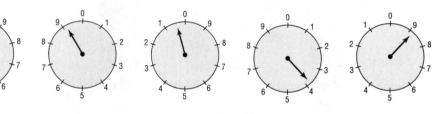

_____

65.

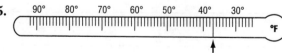

_____

66.

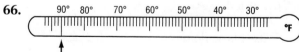

_____

Customary Units of Capacity

What units do you use when you want to find out how much something holds? When you use gallons, cups, and teaspoons you are using customary units of capacity. A teaspoon is a very small customary unit of capacity, and a gallon is a large unit. These are the relationships among customary units of capacity.

1 gallon (gal) = 4 quarts (qt)	1 gal = 16 c
1 qt = 2 pints (pt)	1 c = 16 tablespoons (tbsp)
1 pt = 2 cups (c)	1 tbsp = 3 teaspoons (tsp)

Here are some examples of approximate capacity.

dose of cough medicine
1 tablespoon

serving of coffee
1 cup

container of ice cream
one gallon

container of sour cream
1 pint

container of milk
1 quart

Name three things that are measured, served, or packaged in the given unit.

1. gallons _____

2. cups _____

3. tablespoons _____

Complete each measurement by writing tsp, tbsp, c, pt, qt, or gal.

4. a bottle of cola   2 _____

5. a serving of hot tea   1 _____

6. oil for a car   5 _____

7. a jar of peanut butter   1 _____

8. a dose of medicine $1\frac{1}{2}$ _____

9. flour to make a cake   3 _____

10. water to fill a rain barrel   25 _____

11. salt in a cookie recipe $\frac{1}{2}$ _____

For each exercise, circle the largest capacity and underline the smallest capacity.

12. oil for brownies, oil for a lawnmower, oil for a car

13. a picnic jug, a swimming pool, a child's wading pool

14. a gallon of milk, a cup of water, a pint of juice

15. a can of peaches, a bowl of punch, a bottle of vanilla

16. flour to make cookies, salt to make cookies, oil to make cookies

Here are some common containers used to measure capacity. Some are used to measure very small amounts, and others are used to measure larger amounts. Which of these have you used?

| EXAMPLE 1 | Many people have used a measuring cup. Reading the scale is like reading a ruler. The scale is labeled in cups. Each cup is divided into 4 parts, and each part equals 4 tablespoons. |

Kai tests recipes and writes on-line reviews. To make a recipe he poured rice into the container as shown. How much rice did he pour?

STEP 1 Decide which units to use for the measurement.

_____ c and _____ tbsp

STEP 2 The amount of rice is between 1 and 2 cups.

1 c and _____ tbsp

STEP 3 Find the number of tablespoons more than 1 cup.

1 c and 12 tbsp

ANSWER: Kai poured **1 cup and 12 tablespoons** of rice.

Write the number of cups and tablespoons shown in each container.

17.

_____

18.

_____

19.

_____

Estimating Capacity

Here are two tools used to measure capacity. One is used for very small amounts, and the other is used for larger amounts.

This measuring cup can be used for doses of medicine or other small amounts. It uses teaspoons as a basic unit.

This 2-quart container uses cups as a basic unit.

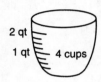

EXAMPLE 1 Ellie runs a small day care center in her home. She used some juice concentrate to make frozen treats for the children. She has $\frac{2}{3}$ cup of the concentrate left over and wants to use it to make juice. The juice directions say to fill the can four times with water and add that much water to a full can of concentrate. Estimate how many quarts of juice Ellie can make from $\frac{2}{3}$ cup of the concentrate. Then mark your estimate on the container.

STEP 1 Calculate the total number of cups.

$$\frac{2}{3} \text{ cup} + 4\left(\frac{2}{3} \text{ cup}\right) = \frac{2}{3} + \frac{8}{3}$$
$$= \frac{10}{3}$$
$$= 3\frac{1}{3} \text{ cups}$$

STEP 2 Estimate the number of quarts in $3\frac{1}{3}$ cups.

$$4 c = 1 \text{ qt, so } 3\frac{1}{3} c \approx \frac{3}{4} \text{ qt}$$

ANSWER:

Estimate the capacity of each object. Draw an appropriate measuring container and mark your estimate on the container.

1. a can of evaporated milk

5. a dose of cough syrup

9. water to make a cake mix

2. a serving of tomato juice

6. amount of coffee in a thermos bottle

10. buttermilk to use in a double batch of muffins

3. amount of water to cook spaghetti

7. amount in a giant-size soft drink

11. amount of canned tomatoes to make chili

4. carbonated water for a punch recipe

8. amount of medicine in an eye dropper

12. one serving of soup

Changing Among Customary Units of Capacity

When you use a smaller unit to measure an amount of liquid, the number of units in your measurement will be larger. When you use a larger unit, the number of units in your measurement will be smaller.

Here are the relations among teaspoons, tablespoons, cups, pints, quarts, and gallons.

1 gallon (gal) = 4 quarts (qt)	1 fluid ounce (fl oz) = 2 tbsp
1 qt = 2 pints (pt)	1 tbsp = 3 tsp
1 pt = 2 cups (c)	1 c = 8 fl oz
	1 pt = 16 fl oz

EXAMPLE 1 Gail, a small engine mechanic, needs 2 quarts of oil to complete her work for the day. She has three 1-pint containers of oil. Does she have enough?

STEP 1 Write the problem. $2 \, qt = \underline{\hspace{1cm}} \, pt$

STEP 2 Write the relationship between pints and quarts. $1 \, qt = 2 \, pt$

STEP 3 Multiply each side of the equation in Step 2 by 2. $2 \, qt = 4 \, pt$

ANSWER: Gail has only 3 pints, but needs **4 pints**. So, she does not have enough.

EXAMPLE 2 Find the number of fluid ounces in 2 cups.

STEP 1 Write the problem. $2 \, c = \underline{\hspace{1cm}} \, fl \, oz$

STEP 2 Write the relationship between cups and fluid ounces. $1 \, c = 8 \, fl \, oz$

STEP 3 Multiply both sides of the equation in Step 2 by 2. $2 \, c = 16 \, fl \, oz$

ANSWER: There are **16 fluid ounces** in 2 cups.

Rewrite each measurement so it uses the given unit.

1. 6 tbsp = _____ tsp   **4.** 2 pt = _____ fl oz   **7.** 20 qt = _____ pt

2. 10 gal = _____ pt   **5.** 4 gal = _____ qt   **8.** 12 fl oz = _____ tbsp

3. 3 qt = _____ c   **6.** 1 c = _____ tbsp   **9.** 25 pt = _____ c

Sometimes a measurement contains a mixed number. Here is an example.

EXAMPLE 3   Change $5\frac{1}{2}$ gallons to quarts.

STEP 1   Write the problem.

$$5\frac{1}{2} \text{ gal} = \underline{\hspace{1cm}} \text{ qt}$$

STEP 2   Write the relationship between gallons and quarts.

$$1 \text{ gal} = 4 \text{ qt}$$

STEP 3   Multiply both sides of the equation in Step 2 by $5\frac{1}{2}$.

$$5\frac{1}{2} \text{ gal} = 22 \text{ qt}$$

ANSWER: $5\frac{1}{2}$ gal = **22 qt**

Rewrite each measurement so it uses the given unit.

10. $1\frac{1}{2}$ c = \underline{\hspace{1cm}} tbsp

11. $3\frac{5}{8}$ gal = \underline{\hspace{1cm}} pt

12. $4\frac{1}{3}$ qt = \underline{\hspace{1cm}} c

13. $2\frac{1}{4}$ tbsp = \underline{\hspace{1cm}} tsp

14. 2.5 fl oz = \underline{\hspace{1cm}} tsp

15. $4\frac{3}{4}$ pt = \underline{\hspace{1cm}} fl oz

16. 6.6 gal = \underline{\hspace{1cm}} qt

17. $7\frac{1}{2}$ pt = \underline{\hspace{1cm}} c

In the next two examples, smaller units are replaced with larger units.

EXAMPLE 4   Tonya is a chef. She plans to make a recipe that calls for 6 cups of cream. At the store she finds that the cream is sold in pint containers. How many pints of cream should she buy?

STEP 1   Write the problem.

$$6 \text{ c} = \underline{\hspace{1cm}} \text{ pt}$$

STEP 2   Write the relationship between cups and pints.

$$2 \text{ c} = 1 \text{ pt}$$

STEP 3   Multiply both sides of the equation in Step 2 by 3.

$$6 \text{ c} = 3 \text{ pt}$$

ANSWER: She should buy **3 pints** of cream.

EXAMPLE 5   Change 9 teaspoons to tablespoons.

STEP 1   Write the problem.

$$9 \text{ tsp} = \underline{\hspace{1cm}} \text{ tbsp}$$

STEP 2   Write the relationship between teaspoons and tablespoons.

$$3 \text{ tsp} = 1 \text{ tbsp}$$

STEP 3   Multiply both sides of the equation in Step 2 by 3.

$$9 \text{ tsp} = 3 \text{ tbsp}$$

ANSWER: Nine teaspoons is the same as **three tablespoons**.

Rewrite each measurement so it uses the given unit.

18. $2 \text{ c} = \underline{\hspace{1cm}} \text{ qt}$

21. $6 \text{ qt} = \underline{\hspace{1cm}} \text{ gal}$

24. $32 \text{ fl oz} = \underline{\hspace{1cm}} \text{ pt}$

19. $32 \text{ tbsp} = \underline{\hspace{1cm}} \text{ c}$

22. $27 \text{ tsp} = \underline{\hspace{1cm}} \text{ tbsp}$

25. $160 \text{ c} = \underline{\hspace{1cm}} \text{ qt}$

20. $10 \text{ c} = \underline{\hspace{1cm}} \text{ pt}$

23. $8 \text{ pt} = \underline{\hspace{1cm}} \text{ gal}$

26. $9 \text{ tsp} = \underline{\hspace{1cm}} \text{ tbsp}$

Sometimes you need to use a mixed number for the smaller units. Here is an example.

EXAMPLE 6   Change 7 cups to pints.

STEP 1   Write the problem.

$$7 \text{ c} = \underline{\hspace{1cm}} \text{ pt}$$

STEP 2   Write the relationship between cups and pints.

$$2 \text{ c} = 1 \text{ pt}$$

STEP 3   Multiply both sides of the equation in Step 2 by $3\frac{1}{2}$.

$$7 \text{ c} = 3\frac{1}{2} \text{ pt}$$

ANSWER: $7 \text{ c} = 3\frac{1}{2} \text{ pt}$

Rewrite each measurement so it uses the given unit.

27. $8 \text{ fl oz} = \underline{\hspace{1cm}} \text{ pt}$

30. $4 \text{ tsp} = \underline{\hspace{1cm}} \text{ tbsp}$

33. $7 \text{ c} = \underline{\hspace{1cm}} \text{ qt}$

28. $22 \text{ qt} = \underline{\hspace{1cm}} \text{ gal}$

31. $30 \text{ fl oz} = \underline{\hspace{1cm}} \text{ pt}$

34. $12 \text{ fl oz} = \underline{\hspace{1cm}} \text{ c}$

29. $20 \text{ tsp} = \underline{\hspace{1cm}} \text{ tbsp}$

32. $16 \text{ pt} = \underline{\hspace{1cm}} \text{ gal}$

35. $25 \text{ tbsp} = \underline{\hspace{1cm}} \text{ c}$

Solve each problem.

36. Mary Jo needs to give medicine to her baby. Each dose is half a teaspoon. Mary Jo has one ounce of the medicine. How many doses are in one ounce?

37. Jose is a produce manager at a supermarket. He wants to display 120 quarts of strawberries. How many gallons of strawberries does he need?

38. Tom manages a gourmet coffee shop. He needs to buy a new coffee maker. He is considering one that makes 8 quarts. How many cups of coffee will it make?

39. Pat is a food preparation worker in a company cafeteria. He needs 3 pints of cream for a recipe. His only clean measuring container is one cup. How many cups does he need?

40. Pat's recipe also calls for 3 tablespoons of sugar. He cannot find a tablespoon, but he has a teaspoon. How many teaspoons of sugar does he need?

41. The soap dispenser in a restaurant kitchen dishwasher holds 4 fluid ounces of soap. If a worker uses a tablespoon to fill the dispenser, how many tablespoons does he need?

42. Jack works in a greenhouse. He has a 10-gallon pot full of soil that he will use for planting. How many shrubs can he plant if he needs one quart of soil for each shrub?

43. Jack uses all the soil. Then he puts more soil in the 10-gallon pot, but only makes it half full. How many flowers can he plant if he needs one pint of soil for each flower?

44. Wendy is a personal trainer. She wants her client to drink 32 fluid ounces of water during a workout. How many cups of water should she tell her client to drink?

45. David is a veterinary assistant. He tells Amy to feed her dog 1 cup of special dog food each day. David wants to give Amy a four-day supply of the dog food. He has a container marked in pints. How many pints of dog food should he give Amy?

46. How many quarts are equal to 20 pints?

47. How many tablespoons are equal to 6 teaspoons?

Estimating with Metric Units of Capacity

In the metric system there are three basic units of capacity. They are **liters, milliliters,** and **kiloliters**. Here are the relationships among these units.

> 1 liter (L) = 1,000 milliliters (mL)
>
> 1 kiloliter (kL) = 1,000 L

In the metric system of measurement, the prefix "milli-" means $\frac{1}{1,000}$ or one-thousandth. The prefix "kilo-" means 1,000. Small amounts are measured in milliliters, and large amounts are measured in kiloliters.

10 mL

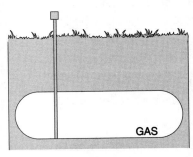

GAS

10 milliliters
a dose of medicine

4 liters
a watering can

2 kiloliters
an underground gas tank

Name three things that are measured in the given unit.

1. milliliters

_____

2. liters

_____

3. kiloliters

_____

Circle the most appropriate unit for measuring the capacity of each item.

4. a sample of blood   mL   L   kL

5. water in a swimming pool   mL   L   kL

6. oil for a can   mL   L   kL

7. amount a tanker holds   mL   L   kL

8. amount a picnic jug holds   mL   L   kL

9. water for a hot tub   mL   L   kL

10. shot of antibiotic   mL   L   kL

11. a keg of beer   mL   L   kL

12. oil to fry an egg   mL   L   kL

Underline the better estimate for the capacity of each object.

13. juice in an orange   16 L   16 mL

14. coffee in a thermos   1 L   1 kL

15. water in a bathtub   28 L   28 kL

16. bleach for 16 loads of laundry   2 kL   2 L

17. salt in a cake mix   3 L   3 mL

18. oil for a car   2.5 kL   2.5 L

19. water in a kitchen sink   4 mL   4 L

20. water to make fruit punch   2 mL   2 L

21. oil from an oil well   6 kL   6 mL

22. water used to take a shower   16 L   16 kL

23. lava from a volcano   206 L   206 kL

24. punch at a wedding reception   16 L   16 kL

25. juice in a can of pineapple   240 L   240 mL

26. blood for a cholesterol test   25 mL   25 L

27. water in a fountain   14 mL   14 kL

Answer each exercise.

28. Circle the smallest unit.   1 kL   1 L   1 mL

29. Circle the largest unit.   1 kL   1 L   1 mL

30. What are the three basic metric units for capacity?

Changing Among Milliliters, Liters, and Kiloliters

Here is a review of the relationships among milliliters, liters, and kiloliters. They are the three basic metric units for capacity.

$$1 \text{ L} = 1,000 \text{ mL}$$
$$1 \text{ kL} = 1,000 \text{ L}$$

EXAMPLE 1 A school nurse wants to give $\frac{1}{4}$ liter of a sports drink along with water to a student who is dehydrated. She has a container of the sports drink labeled in milliliters. How many milliliters of the sports drink should she give the student?

STEP 1 Write the problem.

$$\frac{1}{4} \text{ L} = \underline{\hspace{1cm}} \text{ mL}$$

STEP 2 Write the relationship between milliliters and liters.

$$1 \text{ L} = 1,000 \text{ mL}$$

STEP 3 Multiply each side of the equation in Step 2 by $\frac{1}{4}$.

$$\frac{1}{4} \text{ L} = 250 \text{ mL}$$

ANSWER: The nurse should give the student **250 milliliters** of the sports drink.

EXAMPLE 2 Change 2 liters to kiloliters.

STEP 1 Write the problem.

$$2 \text{ L} = \underline{\hspace{1cm}} \text{ kL}$$

STEP 2 Write the relationship between liters and kiloliters.

$$1,000 \text{ L} = 1 \text{ kL}$$

STEP 3 Divide both sides of the equation in Step 2 by 500.

$$2 \text{ L} = \frac{1}{500} \text{ kL}$$
$$= 0.002 \text{ kL}$$

ANSWER: 2 liters = **0.002 kiloliters**

Rewrite each measurement so it uses the given unit.

1. $4 \text{ L} = \underline{\hspace{1cm}} \text{ mL}$

2. $3 \text{ L} = \underline{\hspace{1cm}} \text{ kL}$

3. $600 \text{ mL} = \underline{\hspace{1cm}} \text{ L}$

4. $10 \text{ L} = \underline{\hspace{1cm}} \text{ kL}$

5. $6 \text{ L} = \underline{\hspace{1cm}} \text{ mL}$

6. $5 \text{ kL} = \underline{\hspace{1cm}} \text{ L}$

Sometimes the given metric measurement has a decimal point. Here is an example.

EXAMPLE 3   Change 4.5 liters to milliliters.

STEP 1   Write the problem. $4.5 \text{ L} = \underline{\hspace{1cm}} \text{ mL}$

STEP 2   Write the relationship between liters and milliliters. $1 \text{ L} = 1{,}000 \text{ mL}$

STEP 3   Multiply both sides of the equation in Step 2 by 4.5. $4.5 \text{ L} = 4{,}500 \text{ mL}$

ANSWER: 4.5 liters = **4,500 milliliters**

Rewrite each measurement so it uses the given unit.

7. $1.5 \text{ L} = \underline{\hspace{1cm}} \text{ mL}$ 10. $3.2 \text{ L} = \underline{\hspace{1cm}} \text{ kL}$

8. $2{,}400 \text{ mL} = \underline{\hspace{1cm}} \text{ L}$ 11. $3.2 \text{ kL} = \underline{\hspace{1cm}} \text{ L}$

9. $5.5 \text{ L} = \underline{\hspace{1cm}} \text{ mL}$ 12. $0.05 \text{ kL} = \underline{\hspace{1cm}} \text{ L}$

Sometimes you need to use a decimal point with a larger unit. Here is an example.

EXAMPLE 4   Change 350 milliliters to liters.

STEP 1   Write the problem. $350 \text{ mL} = \underline{\hspace{1cm}} \text{ L}$

STEP 2   Write the relationship between milliliters and liters. $1{,}000 \text{ mL} = 1 \text{ L}$

STEP 3   Set up a proportion. Use "?" to represent the unknown amount. $\dfrac{350}{1{,}000} = \dfrac{?}{1}$

STEP 4   Find the cross-products and solve for the unknown number.

$$? \times 1{,}000 = 350$$
$$? = \frac{350}{1{,}000} = 0.35$$

ANSWER: 350 milliliters = **0.35 liters**

Here are two more examples of changing a measurement so it uses a larger unit.

EXAMPLE 5 Patrice is a certified nursing assistant in a nursing home. She wants each patient to drink 2,000 milliliters of water every day. How many liters of water is that?

STEP 1 Write the problem.

$$2,000 \text{ mL} = \underline{\hspace{1cm}} \text{ L}$$

STEP 2 Write the relationship between milliliters and liters.

$$1,000 \text{ mL} = 1 \text{ L}$$

STEP 3 Multiply each side of the equation in Step 2 by 2.

$$2,000 \text{ mL} = 2 \text{ L}$$

ANSWER: Patrice wants each patient to drink **2 liters** of water every day.

EXAMPLE 6 Change 400 milliliters to kiloliters.

STEP 1 Find the relationship between milliliters and kiloliters. Each kiloliter is 1,000 liters, and each liter is 1,000 milliliters. In each kiloliter there are 1,000 × 1,000 = 1,000,000 milliliters.

$$1 \text{ kL} = 1,000 \text{ L}$$
$$1 \text{ L} = 1,000 \text{ mL}$$
$$\text{So } 1 \text{ kL} = 1,000,000 \text{ mL}$$

STEP 2 Decide how to convert milliliters to kiloliters.

To change kL to mL, you multiply by 1,000,000. So to change from mL to kL, you divide by 1,000,000.

STEP 3 Convert 400 milliliters to kiloliters.

$$\frac{400}{1,000,000} = \frac{4}{10,000} = 0.0004$$

ANSWER: 400 milliliters = **0.0004 kiloliters**

Rewrite each measurement so it uses the given unit.

13. 10,000 mL = _____ L

14. 600 mL = _____ kL

15. 150 mL = _____ L

16. 780 mL = _____ kL

17. 100,000 mL = _____ L

18. 800 mL = _____ kL

19. 6,200 mL = _____ L

20. 9,225 mL = _____ kL

Adding and Subtracting with Capacity Measurements

Sometimes you need to add or subtract measurements of capacity.

EXAMPLE 1 Sam, a painter, used 3 gallons and 3 quarts of paint for a morning job. He used 2 gallons and 3 quarts for an afternoon job. How much paint did he use in all?

STEP 1 Line up the measurements, putting like units under like units.

$$\begin{array}{r} 3\ gal\ 3\ qt \\ +\ 2\ gal\ 3\ qt \\ \hline \end{array}$$

STEP 2 Add the quarts and add the gallons.

$$\begin{array}{r} 3\ gal\ 3\ qt \\ +\ 2\ gal\ 3\ qt \\ \hline 5\ gal\ 6\ qt \end{array}$$

STEP 3 Think of 6 quarts as 4 quarts + 2 quarts. So 6 quarts = 1 gallon 2 quarts. Rewrite the sum.

$$5\ gal + 6\ qt = 5\ gal + 4\ qt + 2\ qt$$
$$= 5\ gal + 1\ gal + 2\ qt$$
$$= 6\ gal\ 2\ qt$$

ANSWER: Sam used **6 gallons and 2 quarts** of paint in all.

EXAMPLE 2 Subtract 3 pints and $1\frac{1}{2}$ cups from 5 pints and 1 cup.

STEP 1 Line up the measurements, putting like units under like units.

$$\begin{array}{r} 5\ pt\ 1\ c \\ -\ 3\ pt\ 1\frac{1}{2}\ c \\ \hline \end{array}$$

STEP 2 "Borrow" by changing 5 pints to 4 pints + 2 cups. So, rewrite the top number as 4 pints + 3 cups.

$$\begin{array}{r} 4\ pt\ 3\ c \\ -\ 3\ pt\ 1\frac{1}{2}\ c \\ \hline \end{array}$$

STEP 3 Subtract the pints and subtract the cups.

$$\begin{array}{r} 4\ pt\ 3\ c \\ -\ 3\ pt\ 1\frac{1}{2}\ c \\ \hline 1\ pt\ 1\frac{1}{2}\ c \end{array}$$

ANSWER: The difference is **1 pint $1\frac{1}{2}$ cups**.

Add or subtract.

1. $\begin{array}{r} 6\ gal\ 3\ qt \\ +\ 2\ gal\ 2\ qt \\ \hline \end{array}$

2. $\begin{array}{r} 6\ c\ 3\ tbsp \\ -\ 2\ c\ 6\ tbsp \\ \hline \end{array}$

3. $\begin{array}{r} 12\ gal\ 13\ c \\ +\ 4\ gal\ \ \ 8\ c \\ \hline \end{array}$

4. $\begin{array}{r} 2\ gal\ 1\ qt \\ +\ 3\ gal\ 1\ qt \\ \hline \end{array}$

5. $\begin{array}{r} 2\ qt\ 3\ c \\ +\ 6\ qt\ 2\ c \\ \hline \end{array}$

6. $\begin{array}{r} 7\ pt\ 1\ c \\ -\ 3\ pt\ 2\ c \\ \hline \end{array}$

7. $\begin{array}{r} 6\ pt\ 0\ c \\ -\ 3\ pt\ 1\ c \\ \hline \end{array}$

8. $\begin{array}{r} 2\ qt\ 3\ pt \\ +\ 6\ qt\ 2\ pt \\ \hline \end{array}$

9. $\begin{array}{r} 6\ qt\ 1\ c \\ -\ 4\ qt\ 3\ c \\ \hline \end{array}$

10. $\begin{array}{r} 2\ pt\ 1\ c \\ +\ 6\ pt\ 1\ c \\ \hline \end{array}$

11. $\begin{array}{r} 4\ qt\ 2\ c \\ -\ 1\ qt\ 3\ c \\ \hline \end{array}$

12. $\begin{array}{r} 4\ gal\ 0\ qt\ 0\ pt \\ -\ \ \ \ \ \ \ \ \ 2\ qt\ 1\ pt \\ \hline \end{array}$

To add or subtract metric measurements, be sure to use the same unit.

EXAMPLE 3   Find the sum and the difference of 5.2 L and 1,350 mL.

STEP 1   Rewrite one of the measurements
so both use the same units.
This example uses milliliters.

5.2 L = _____ mL
1 L = 1,000 mL
5.2 L = 5,200 mL

STEP 2   Line up the two measurements.
Then add or subtract.

Add	Subtract
5,200 mL	5,200 mL
+ 1,350 mL	− 1,350 mL
6,550 mL	3,850 mL

ANSWER: The sum is 6,550 milliliters. The difference is 3,850 milliliters.

Find each sum or difference.

13. 35 L + 2 kL

14. 3 L − 2,540 mL

15. 7.2 L + 3,500 L

16. 8.65 L − 1,400 mL

17. 2.35 L + 450 mL

18. 2.6 kL − 1,750 L

Solve each problem.

19. If Juan starts with 2 gallons 3 quarts and adds 2 gallons 2 quarts, how much total liquid does he have?

20. Sandra is a laboratory technician at a chemical company. She took 750 milliliters of ethylene glycol from a full 2.5-liter container to run an experiment. She needed to record how much was left in the container. How much was left?

21. Sam works for a caterer. His boss tells him to add punch to a bowl that already contains $6\frac{1}{2}$ liters of punch. Sam adds 2.5 liters. His boss asks him how much punch is in the bowl now. How much punch is in the bowl?

22. A recipe made 1 gallon 5 cups of soup. Jay, a server, serves 16 cups of the soup one day. He needs to know how much is left for the next day. How much soup is left?

Multiplying and Dividing with Capacity Measurements

Sometimes you need to multiply or divide measurements of capacity. For example, to find the total capacity of several same-size containers, you can multiply. To find out how many times you can fill a small container from a large one, you can divide.

EXAMPLE 1 Digna is an automotive service manager. She buys car oil in 15-gallon drums. If the average car uses $1\frac{1}{2}$ gallons of oil, how many cars can she have serviced with one 15-gallon container of oil?

STEP 1 Decide which math operation to use.

She wants to know how many times a large container (the 15-gallon drum) can be used to fill smaller containers (the cars). The operation is division.

STEP 2 Write the problem.

$$15 \div 1\frac{1}{2} = 15 \div \frac{3}{2}$$

STEP 3 Divide 15 by $\frac{3}{2}$. That is the same as multiplying 15 by $\frac{2}{3}$.

$$= \frac{15}{1} \times \frac{2}{3}$$

STEP 4 Multiply the two fractions.

$$= \frac{15 \times 2}{1 \times 3} = \frac{30}{3} = 10$$

ANSWER: Digna can have **10 cars** serviced with one 15-gallon drum of oil.

EXAMPLE 2 Multiply 6 quarts 2 cups by 5.

STEP 1 Write the problem.

$$\begin{array}{r} 6\ qt\ 2\ c \\ \times \quad\ \ 5 \\ \hline \end{array}$$

STEP 2 Multiply 5 times the number of cups and multiply 5 times the number of quarts.

$$\begin{array}{r} 6\ qt\ 2\ c \\ \times \quad\ \ 5 \\ \hline 30\ qt\ 10\ c \end{array}$$

STEP 3 One quart is 4 cups, so replace 10 cups with 2 quarts and 2 cups.

$$30\ qt\ 10\ c = 30\ qt + 2\ qt + 2\ c$$
$$= 32\ qt\ 2\ c$$

ANSWER: The product is **32 quarts 2 cups**.

EXAMPLE 3 Divide 4 liters and 800 milliliters by 12.

STEP 1 Use the same unit for both measurements. Using liters, 800 milliliters = 0.8 liters.

$$4\,L + 800\,mL = 4\,L + 0.8\,L$$
$$= 4.8\,L$$

STEP 2 Write the problem.

$$\frac{4.8}{12} = ?$$

STEP 3 Divide.

$$\frac{4.8}{12} = 0.4$$

ANSWER: The quotient is **0.4 liters**.

Multiply or divide.

1. 2 pt 1 c × 6

2. 4 gal 2 qt ÷ 2

3. 5 L 250 mL ÷ 5

4. 3 kL 500 L ÷ 2

5. 3 gal 3 qt × 6

6. 3 qt 1 pt × 10

7. 6 gal 2 qt ÷ 4

8. 6.4 L × 8

9. 2.7 kL ÷ 3

10. 16 L 200 mL ÷ 4

11. 6 qt 4 c ÷ 4

12. 2 gal 3 pt × 6

13. 12 qt 1 c × 4

14. 6 c 4 tbsp × 3

15. 6 kL 500 mL ÷ 25

Solve each problem.

16. Tami bought 5 cans of evaporated milk. Each has a capacity of 354 milliliters. How many milliliters did she buy? How many liters?

17. Tru offers soup each day at his bakery. How much broth does he need for a double batch of chicken soup if a single batch recipe calls for 1 quart 3 cups of broth?

18. Sami is a laboratory instructor at a community college. She assigns a lab test that requires 80 milliliters of liquid each time the test is run. She tells each student to run the test twelve times. How much liquid does Sami need to purchase for each student?

19. Fay is a hotel housekeeper. She has $2\frac{1}{2}$ quarts of detergent in stock. She uses $\frac{1}{2}$ cup in each load of wash. She needs to know if she has enough to do all the wash for the day. How many loads of wash can she do with the detergent she has in stock?

Comparing Customary and Metric Units of Capacity

Sometimes a measurement of capacity is listed in both metric units and customary units. For example, the wrapper of a large soda container lists the capacity as "2 liters (67.6 fl oz)".

At the right are some comparisons between metric and customary units.

1 L ≈ 1.1 qt (so 1 L > 1 qt)
1 mL ≈ 2 drops
1 kL = 1,000 L ≈ 275 gal

EXAMPLE Eva is grocery shopping for her day care center. She is deciding whether to buy a liter of juice for $1.19 or a quart of juice for $1.19. Which is the better buy?

STEP 1 Write the two units.   1 L   1 qt

STEP 2 Decide which is larger.   1 L > 1 qt

STEP 3 Compare the prices.   The prices are the same.

ANSWER: **One liter is larger than one quart, so one liter for $1.19 is a better buy than one quart for $1.19.**

Circle the larger measurement.

1. 1 qt or 1 mL

2. 1 gal or 1 kL

3. 1 L or 1 c

4. 1 mL or 1 fl oz

5. 1 pt or 1 L

6. 1 gal or 1 L

7. 1 tsp or 1 mL

8. 1 qt or 1 gal

9. 1 kL or 1 L

10. 1 c or 1 pt

11. 1 qt or 1 kL

12. 1 fl oz or 1 tbsp

Circle the more appropriate measurement for the capacity of each object.

13. can of evaporated milk   400 mL or 400 fl oz

14. water in a swimming pool   1,000 kL or 1,000 qt

15. oil in a small truck   6 qt or 6 kL

16. glass of juice   1 L or 1 c

17. gas in a full tank of a sports car   15 L or 15 gal

18. container of bleach in a laundry room   1 gal or 1 kL

19. oil to make brownies   6 fl oz or 6 mL

Focus on Geometry: Volume

Volume (V) is the amount of space inside a three-dimensional figure. Volume is measured in cubic units. For a three-dimensional figure shaped like a box, you can use the formula $V = \ell \times w \times h$ to calculate volume.

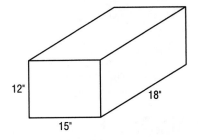

EXAMPLE 1 Carla is a packaging technologist. She needs to know the volume of a packing box that is 18 inches by 15 inches by 12 inches. What should she do to calculate the volume? What is the volume of the box?

STEP 1 Write the formula for volume.

$$V = \ell \times w \times h$$

STEP 2 Substitute the values of ℓ, w, and h into the formula.

$$V = 18 \times 15 \times 12$$

STEP 3 Multiply the numbers.

$$V = 3{,}240$$

ANSWER: She should multiply the three measurements. The volume of the box is **3,240 cubic inches**.

Find the volume of each box. Use the formula $V = \ell \times w \times h$.

1.

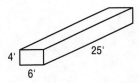

$V =$ _____

3.

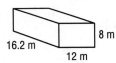

$V =$ _____

2.

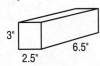

$V =$ _____

4.

$V =$ _____

5. Name three situations in which you might want to calculate volume.

_____

Sometimes the dimensions are not given in the same units. Before you can use the volume formula, all three dimensions must be given in the same unit.

EXAMPLE 2 Find the volume of a box 4 feet long, 2 feet wide, and 9 inches tall.

STEP 1 Use a single unit for all three measurements. To use feet as the unit, change 9 inches to feet.

$$9 \text{ in.} = \frac{9}{12} \text{ ft}$$

STEP 2 Write the formula.

$$V = \ell \times w \times h$$

STEP 3 Substitute the values of ℓ, w, and h into the formula.

$$V = 4 \times 2 \times \frac{3}{4}$$

STEP 4 Multiply.

$$V = 6$$

ANSWER: The volume is **6 cubic feet**.

Find the volume of each box.

6. $\ell = 10$ ft
    $w = 1$ yd
    $h = 5$ ft
    $V = $ _____

7. $\ell = 6$ m
    $w = 500$ mm
    $h = 2$ m
    $V = $ _____

8. $\ell = 3$ yd
    $w = 1$ ft
    $h = 2$ yd
    $V = $ _____

9. $\ell = 2$ in.
    $w = 6$ in.
    $h = 18$ in.
    $V = $ _____

10. $\ell = 4$ yd
    $w = 36$ in.
    $h = 6$ ft
    $V = $ _____

11. $\ell = 350$ mm
    $w = 2$ m
    $h = 0.05$ m
    $V = $ _____

A **cube** is a three-dimensional figure that has squares as all the faces of the figure. If the length of each side of each square is s, then a formula to find the volume of a cube is $V = s^3$.

12. Name some common objects that are cubes.

_____

Find the volume of each cube.

13.

2 yd

$V =$ _____

15.

2.5 m

$V =$ _____

14.

6"

$V =$ _____

16.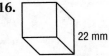

22 mm

$V =$ _____

Solve each problem.

17. What is the volume of a cube whose edges are 15 inches?

18. What is the volume of a rectangular box that is 6 inches by 8 inches by 3 inches?

19. A sugar cube is about a half-inch on each edge. What is the volume of the sugar cube?

20. A builder is designing a self storage facility. He designs some of the storage units so that their inside is in the shape of a cube whose edges are 2.4 meters. He needs to know the volume of the storage space. What is that volume?

If you know the volume of a box and measurements for two of its dimensions, you can use the formula $V = \ell \times w \times h$ to find the measurement for the third dimension.

EXAMPLE 3 The volume of a box is 480 cubic inches. If the height is 10 inches and the width is 6 inches, what is the length of the box?

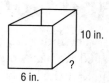

STEP 1 Write the formula.

$$V = \ell \times w \times h$$

STEP 2 Substitute the known values into the formula. Don't forget that the value for V is known.

$$480 = \ell \times 6 \times 10$$

STEP 3 Simplify the expression on the right.

$$480 = 60 \times \ell$$

STEP 4 Divide both sides of the equation in Step 3 by 60.

$$8 = \ell$$

ANSWER: The length of the box is **8 inches**.

Find the missing dimension for each box.

21. $V = 48$ cu ft

6 ft 2 ft ?

22. $V = 1,200$ cu m

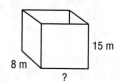

8 m 15 m ?

23. $V = 800$ cu in.

25 in. 4 in. ?

Solve each problem.

24. The volume of a box is 1,200 cubic inches. The width is 6 inches and the height is 10 inches. What is the length of the box?

25. The volume of a cube is 1,000 cubic inches. How long is each edge?

26. Tina, a package designer, is designing a cereal box. Based on marketing research, she wants to make the box 1 foot high and 4 inches thick. If the volume of the box is to be 576 cubic inches, what should be the length of the box?

27. If the area of one face of a cube is 64 square inches, what is its volume?

28. Ray, a concrete worker, is making a sidewalk 2 feet wide and 6 inches deep. He has 3 cubic feet of concrete mixed. What length of sidewalk can he make with the mixed concrete?

Focus on Geometry: Volume of Cylinders and Cones

A **cylinder** is a can-shaped container. The top and bottom, which are circles, are called the **bases** of the cylinder. To calculate the volume of a cylinder, you can multiply the height times the area of the base.

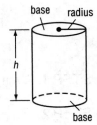
base   radius

h

base

Volume of a cylinder = area of base × height
$V = B \times h$ where B is the area of the base
$V = \pi r^2 h$ where $\pi \approx 3.14$ and r is the radius of the base

EXAMPLE 1 Freeda provides gardening services to clients. She wants to plant a shrub in a cylindrical planter that is 2 feet tall and has a radius of 3 feet. She needs the volume of the planter so she can figure out how much dirt and compost to use. What is the volume of the planter?

STEP 1	Write the formula.	$V = \pi r^2 h$
STEP 2	Substitute the values for π, r, and h into the formula.	$V = (3.14)(3^2)(2)$
STEP 3	Evaluate 3^2.	$3^2 = 3 \times 3 = 9$
STEP 4	Multiply.	$V = (3.14)(9)(2)$ $= 56.52$

ANSWER: The volume of the planter is about **56.5 cubic feet**.

Find the volume of each cylinder.

1.

2'

12'

$V = _____$

3.
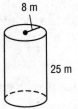
8 m

25 m

$V = _____$

5.

2.5 cm

7.5 cm

$V = _____$

2.
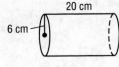
20 cm

6 cm

$V = _____$

4.
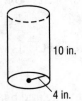
10 in.

4 in.

$V = _____$

6.
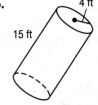
4 ft

15 ft

$V = _____$

Sometimes the dimensions for a cylinder are not given in the same units. Before you can use the volume formula, all the dimensions must use the same unit.

EXAMPLE 2   Find the volume of a cylinder that is 4 feet high and has a radius of 18 inches.

STEP 1   Use a single unit for all the measurements. To use feet as the unit, change 18 inches to feet.

$$18 \text{ in.} = \frac{18}{12} \text{ ft}$$
$$= 1\frac{1}{2} \text{ ft}$$

STEP 2   Write the formula.

$$V = \pi r^2 h$$

STEP 3   Substitute the values for π, r, and h into the formula.

$$V = (3.14)\left(\tfrac{3}{2}\right)\left(\tfrac{3}{2}\right)(4)$$

STEP 4   Multiply the numbers.

$$V = 28.26 \text{ cu ft}$$

ANSWER: The volume of the cylinder is **28.26 cubic feet**.

Find the volume of each cylinder.

7.   4 cm

120 mm

$V = $ _____

8.
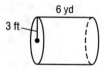
3 ft   6 yd

$V = $ _____

9.   2 in.

7 ft

$V = $ _____

10.
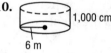
1,000 cm

6 m

$V = $ _____

     If you are given the diameter of a cylinder, you can find the radius because the diameter is twice the radius.

EXAMPLE 3   Find the volume of a cylinder that is 10 feet high and has a diameter of 6 feet.

STEP 1   Find the value of the radius.

$$r = \tfrac{1}{2}d = \left(\tfrac{1}{2}\right)(6) = 3$$

STEP 2   Write the formula.

$$V = \pi r^2 h$$

STEP 3   Substitute the values of π, r, and h into the formula. Then multiply.

$$V = (3.14)(3)(3)10$$
$$= 282.6$$

ANSWER: The volume is **282.6** cubic feet.

Find the volume of each cylinder.

11.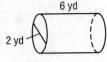

6 yd

2 yd

$V =$ _____

13.

20'

16'

$V =$ _____

12.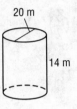

20 m

14 m

$V =$ _____

14.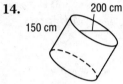

200 cm

150 cm

$V =$ _____

A **cone** is a container that has a circular **base** at one end and a point at the other end. The **height** of a cone is the distance from the center of the circular base to the point.

Here is the formula for the volume of a cone:

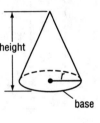

height

base

$V = \frac{1}{3}\pi r^2 h$ where $\pi \approx 3.14$,
     r is the radius of the base,
     and h is the height.

EXAMPLE 4   **Find the volume of a cone with height 6 inches and radius 2 inches.**

STEP 1   Write the formula.

$V = \frac{1}{3}\pi r^2 h$

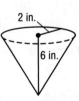

2 in.

6 in.

STEP 2   Substitute values for π, r, and h into the formula.

$V = \frac{1}{3}(3.14)(2^2)(6)$

$\quad = \frac{1}{3}(3.14)(2 \times 2)(6)$

STEP 3   Multiply.

$V = 25.12$

ANSWER: The volume of the cone is **25.12 cubic feet**.

Find the volume of each cone.

15.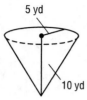

5 yd

10 yd

$V = $ _____

18.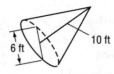

9 in.

18 in.

$V = $ _____

16.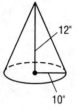

12"

10"

$V = $ _____

19.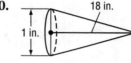

6 ft

10 ft

$V = $ _____

17. 8 cm

12 cm

$V = $ _____

20.

18 in.

1 in.

$V = $ _____

Solve each problem.

21. An engineer is designing a cylindrical underground tank for a gas station. She makes the tank 30 feet long with a diameter of 18 feet. She needs to know how much gasoline the tank will hold. What is the volume of the tank? If a cubic foot is about 7.5 gallons, approximately how many gallons of gasoline does the tank hold?

22. A cylindrical gas tank is 6 feet long and has a diameter of 5 feet. What is the volume of the tank?

23. A cylinder is 10 inches tall. If the volume of the cylinder is 785 cubic inches, what is the radius of the cylinder?

24. If the volume of a cone with a 6-centimeter diameter is 282.6 cubic centimeters, what is the height of the cone?

25. What is the volume of a cone that is 6 inches tall and has a radius of 4 inches?

Section 4 Cumulative Review

Rewrite each measurement so it uses the given unit.

1. 1 qt = _____ c

2. 2 pt = _____ qt

3. 1 L = _____ mL

4. _____ tsp = 1 tbsp

5. _____ fl oz = 1 tbsp

6. 1 c = _____ fl oz

7. 1 L = _____ kL

8. 1 qt = _____ fl oz

9. _____ qt = 1 gal

Mark the measurement on each container.

10. 75 mL

11. $1\frac{1}{2}$ fl oz

12. $1\frac{3}{4}$ c

Circle the smaller measurement.

13. 1 L or 1 qt

14. 1 mL or 1 fl oz

15. 1 c or 1 qt

16. 1 L or 1 gal

17. 1 tsp or 1 c

18. 1 pt or 1 L

Underline the best estimate for the capacity of each object.

19. a dose of medicine   2 fl oz, 2 c, 2 pt

20. a serving of ice cream   8 mL, 8 c, 8 fl oz

21. a serving of coffee   1 qt, 1 mL, 1 c

22. oil for a lawn mower   1 gal, 1 qt, 1 tbsp

23. water for a punch recipe   1 kL, 1 L, 1 mL

Rewrite each measurement so it uses the given unit.

24. 6 c = _____ pt

25. _____ qt = 42 c

26. 4 gal = _____ qt

27. 6 fl oz = _____ tbsp

28. 750 mL = _____ L

29. 2.4 kL = _____ L

30. 18 qt = _____ gal

31. 28 qt = _____ c

32. _____ mL = 1.6 L

Write the abbreviation for each unit.

33. fluid ounce _____

34. liter _____

35. cup _____

36. gallon _____

37. milliliter _____

38. quart _____

Rewrite each measurement so it uses the given unit.

39. _____ mL = 2.4 L

40. 3 pt = _____ c

41. $2\frac{1}{2}$ gal = _____ qt

42. _____ fl oz = 3 tbsp

43. 6 gal = _____ pt

44. 18 c = _____ pt

45. 9 tsp = _____ tbsp

46. 15 c = _____ qt

Write the measurements from smallest to largest.

47. 3 c, 3 tsp, 3 pt _____

48. 750 mL, 7 L, 1500 mL _____

49. $6\frac{1}{2}$ qt, 14 qt, 28 c _____

Underline the largest amount.

50. 6 c, 2 qt, 3 pt

51. 2 L, 6,500 mL, 1 kL

52. 1 fl oz, 10 tsp, 4 tbsp

Circle the best estimate for each amount.

53. amount of blood to test cholesterol   50 L   700 mL   250 mL

54. amount of car oil for a sports car   5 gal   5 qt   5 c

55. amount of water to take a shower   15 qt   15 kL   15 gal

Perform each operation.

56.
$$\begin{array}{r} 6 \text{ qt } 2 \text{ c} \\ - 3 \text{ qt } 3 \text{ c} \\ \hline \end{array}$$

57.
$$\frac{6 \text{ gal } 2 \text{ qt}}{4}$$

58.
$$\begin{array}{r} 7 \text{ L } 600 \text{ mL} \\ + \quad 3,450 \text{ mL} \\ \hline \end{array}$$

59.
$$\begin{array}{r} 3 \text{ pt } 1 \text{ c} \\ \times \quad 5 \\ \hline \end{array}$$

Solve each problem.

60. Find the volume of a cube with 6-inch edges.

61. What is the volume of a cylinder that is 10 meters high and has a diameter of 6 meters?

62. Henry is a paving contractor. He wants to make a driveway 40 feet long, 8 feet wide, and 6 inches thick. He needs to know how much concrete to order. What volume does he need?

63. Mark had to add 7 cups of oil to his car. How many quarts is that?

64. A painter needs $1\frac{1}{2}$ gallons of paint to finish a job. The store only has the color she needs in quarts. How many quarts does she need?

65. What is the volume of a rectangular solid that is 4 inches by 5 inches by 6 inches?

66. A recipe calls for $1\frac{1}{2}$ pints of tomatoes. How many cups is that?

67. A doctor tells her patient to take medicine in one-tablespoon doses. The doctor has a 6-fluid ounce sample bottle of the medicine for the patient, and needs to know how many doses are in the bottle. How many doses are in the bottle?

68. The volume of a box is 1,400 cubic inches. The width is 7 inches and the length is 10 inches. What is the height of the box?

69. What is the volume of a cone that is 6 inches high and has a radius of 3 inches?

70. Jason is a licensed vocational nurse in a hospital. He needs to make sure his patient gets a liter of fluid before he finishes his shift. The patient has received 250 milliliters. How much more fluid should the patient get?

List the names of the units.

71. List the customary units and the metric units for measuring weight.

_____

_____

72. List the customary units and the metric units for measuring capacity.

_____

_____

73. List the customary units and the metric units for measuring length.

_____

_____

Write the measurement shown on each ruler.

74. _____

75. _____

76. _____

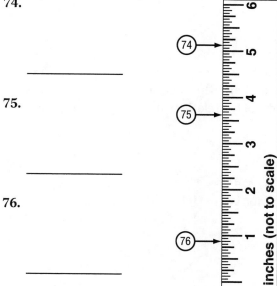

77. _____

78. _____

79. _____

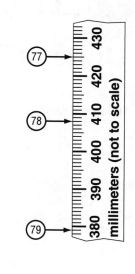

Find each indicated length.

80.

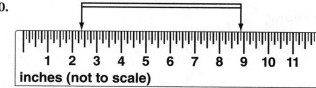

_____

81.

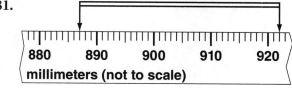

_____

Find the complement of each angle.

82.

 63°

_____

83.

 27°

_____

Find the supplement of each angle.

84.

 115°

_____

85.

 45°

_____

TIME AND VELOCITY

Units of Time

When you measure a brief amount of time, you might use **seconds** or **minutes** as the unit. Other familiar units of time are **hours**, **days**, **weeks**, and **months**. Longer units of time are **years**, **decades**, and **centuries**. Here are some of the relationships among these units of time.

1 minute (min)	=	60 seconds (sec)	1 yr	=	52 weeks
1 hour (hr)	=	60 min	1 decade	=	10 years
1 day	=	24 hr	1 century	=	100 years
1 week	=	7 days			
1 year (yr)	=	12 months (mo)			

Here are some examples of approximate time.

Saying the words "ten thousand" takes about 1 second.

Washing your hands takes about 1 minute.

Baking a loaf of bread takes about 1 hour.

Name three things that are measured in the given unit.

1. seconds

   _____

2. weeks

   _____

3. hours

   _____

Write the unit that is most appropriate to measure each amount of time.

4. length of time to pay off a mortgage _____

5. time needed to wallpaper a kitchen _____

6. time needed to cook a hamburger _____

7. how long it takes to fill a car's gas tank _____

8. how long someone studies to become a doctor _____

9. time on a train between Seattle and San Francisco _____

10. amount of time to eat three pizza slices _____

11. length of time in one school semester _____

Circle the time that is more appropriate for each event.

12. time to fly 1,200 miles

2 days or 2 hours

13. time to work the early shift

8 minutes or 8 hours

14. time to wash dinner dishes

20 minutes or 20 seconds

15. time to grow tomatoes

4 days or 4 weeks

16. time it takes to build a store

15 days or 15 months

17. time it takes to jog 2 miles

20 minutes or 20 hours

18. time it takes to lose 10 pounds

12 weeks or 12 hours

19. time it takes to bake a cake

40 minutes or 40 hours

A stopwatch can be used to measure short periods of time, such as how long a runner takes to complete a race. A period of time is often called an **interval** or a **duration**.

EXAMPLE Janice, a track and field coach, needs to be able to use a stopwatch to time her runners. To time a runner for a 400-meter run, she pressed the button once to start the watch, and then again to stop the watch. According to her stopwatch, how long did it take the runner to run 400 meters?

STEP 1 Decide what units are used.

STEP 2 See where the arrow has stopped.

minutes and seconds

The watch shows 1 minute and 8 seconds.

ANSWER: The runner ran 400 meters in **1 minute and 8 seconds**.

Write the duration shown on each stop watch.

20.

duration: _____

21.

duration: _____

22.

duration: _____

Write the duration shown on each stop watch.

23.

duration: _____

25.

duration: _____

27.

duration: _____

24.

duration: _____

26.

duration: _____

28.

duration: _____

Mark the given duration on the face of the stop watch.

29.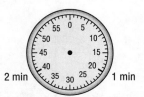

1 min 10 sec

31.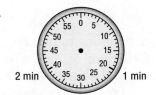

8 seconds

33.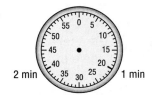

2 min 37 sec

30.

2 min 15 sec

32.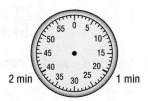

1 min 51 sec

34.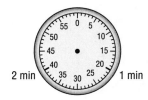

0 min 24 sec

Estimating Time

Estimate how long it would take you to do each activity.

1. walk to work or school

   _____

2. iron a shirt

   _____

3. scrub the kitchen floor

   _____

4. watch a soccer game

   _____

5. take a shower

   _____

6. walk 4 miles

   _____

7. take the trash out

   _____

8. cook a 15-pound turkey

   _____

9. read a magazine

   _____

10. wash a car

    _____

Complete each measurement by writing an appropriate unit of time.

11. vacation after being with a
    company for 10 years

    3 _____

12. lunch break

    30 _____

13. good nights' rest

    8 _____

14. time to get a drink at a water fountain

    20 _____

15. time to buy groceries

    40 _____

16. time for snow drifts to melt

    6 _____

17. time for a movie

    $2\frac{1}{2}$ _____

18. time worked in one week

    45 _____

19. time to earn a nurse's license

    3 _____

20. time of summer break for students

    10 _____

Changing Among Units of Time

When you use a smaller unit to measure a time interval, the number of units in your measurement will be larger. When you use a larger unit, the number of units in your measurement will be smaller.

At the right is a review of the relationships among seconds, minutes, hours, and days.

1 min = 60 sec
1 hr = 60 min
1 day = 24 hr

EXAMPLE 1   Connie writes on-line film reviews. With each review she includes the running time of the film in minutes. What is the running time of a 3-hour film in minutes?

STEP 1   Write the problem.

$$3 \, hr = \underline{\quad} \, min$$

STEP 2   Write the relationship between minutes and hours.

$$1 \, hr = 60 \, min$$

STEP 3   We want to go from 1 hour to 3 hours. So multiply both sides of the equation in Step 2 by 3.

$$3 \, hr = 180 \, min$$

ANSWER: The running time of the film is **180 minutes**.

EXAMPLE 2   How many minutes are there in 4 days?

STEP 1   Write the problem.

$$4 \, days = \underline{\quad} \, min$$

STEP 2   Write the relationship between days and hours. Each hour is the same as 60 minutes, so multiply $24 \times 60 = 1{,}440$ to find the number of minutes in one day.

$$1 \, day = 24 \, hr$$
$$1 \, day = (24)(60) \, min$$
$$1 \, day = 1{,}440 \, min$$

STEP 3   We want to go from 1 day to 4 days, so multiply both sides of the equation "1 day = 1,440 min" by 4.

$$4 \, days = 5{,}760 \, min$$

ANSWER: There are **5,760 minutes** in 4 days.

Rewrite each measurement so it uses the given unit.

1. 120 hr = ____ sec

2. 5 days = ____ sec

3. 72 days = ____ hr

4. 2 days = ____ min

5. 4 hr = ____ min

6. 10 min = ____ sec

7. 7 days = ____ min

8. 3 hr = ____ min

9. 45 min = ____ sec

Sometimes a time measurement contains a mixed number. Here is an example.

EXAMPLE 3   Change $3\frac{1}{2}$ days to hours.

STEP 1   Write the problem.

$$3\frac{1}{2} \text{ days} = \underline{\quad} hr$$

STEP 2   Write the relationship between days and hours.

$$1 \text{ day} = 24 \text{ hr}$$

STEP 3   We want to go from 1 day to $3\frac{1}{2}$ days. So multiply both sides of the equation in Step 2 by $3\frac{1}{2}$.

$$3\frac{1}{2} \text{ days} = 24 \times 3\frac{1}{2}$$
$$= (24)(3) + (24)\left(\frac{1}{2}\right)$$
$$= 72 + 12$$
$$= 84$$

ANSWER: $3\frac{1}{2}$ days = **84 hours**

Rewrite each measurement so it uses the given unit.

10. 2.5 hr = _____ min

11. $6\frac{1}{4}$ days = _____ hr

12. 12.3 min = _____ sec

13. 4.2 hr = _____ min

14. 2.5 days = _____ hr

15. 3.8 min = _____ sec

16. $2\frac{3}{4}$ hr = _____ min

17. $1\frac{1}{2}$ days = _____ min

18. 5.6 min = _____ sec

EXAMPLE 4   Leon is a teacher, giving the GED writing test to a group of students. It is a 120-minute test. Leon needs to know the test time in hours so he can set the start time and end time. What is the test time in hours?

STEP 1   Write the problem.

$$120 \text{ min} = \underline{\quad} hr$$

STEP 2   Write the relationship between minutes and hours.

$$60 \text{ min} = 1 \text{ hr}$$

STEP 3   We want to go from 60 minutes to 120 minutes. So multiply both sides of the equation in Step 2 by 2.

$$120 \text{ min} = 2 \text{ hr}$$

ANSWER: It is a **2-hour** test.

EXAMPLE 5   Change 180 seconds to minutes.

STEP 1   Write the problem.

$$180 \text{ sec} = \underline{\quad} min$$

STEP 2   Write the relationship between seconds and minutes.

$$60 \text{ sec} = 1 \text{ min}$$

STEP 3   We want to go from 60 seconds to 180 seconds. So multiply both sides of the equation in Step 2 by 3.

$$180 \text{ sec} = 3 \text{ min}$$

ANSWER: 180 seconds = **3 minutes**

Rewrite each measurement so it uses the given unit.

19. 3,600 sec = _____ min

22. 168 hr = _____ days

25. 300 sec = _____ min

20. 48 hr = _____ days

23. 600 sec = _____ min

26. 240 min = _____ hr

21. 120 min = _____ hr

24. 96 hr = _____ days

27. 1,440 min = _____ hr

When you change from a smaller unit to a larger unit, the new measurements may contain a mixed number or a decimal. Here are two examples.

`EXAMPLE 6` **Change 105 minutes to hours.**

STEP 1 Write the problem.

$$105\ min = \underline{\quad} hr$$

STEP 2 Write the relationship between minutes and hours.

$$60\ min = 1\ hr$$

STEP 3 Set up a proportion with minutes on the top and hours on the bottom. Use "?" to represent the unknown number of hours.

$$\frac{105}{?} = \frac{60}{1}$$

STEP 4 Find the cross-products and solve for the unknown number.

$$? \times 60 = 105 \times 1$$
$$? = \frac{105}{60} = \frac{7}{4} = 1\frac{3}{4}$$

ANSWER: 105 minutes = $1\frac{3}{4}$ **hours**

`EXAMPLE 7` **Change 204 hours to days.**

STEP 1 Write the problem.

$$204\ hr = \underline{\quad} days$$

STEP 2 Write the relationship between hours and days.

$$24\ hr = 1\ day$$

STEP 3 Set up a proportion with hours on the top and days on the bottom. Use "?" to represent the unknown number of days.

$$\frac{204}{?} = \frac{24}{1}$$

STEP 4 Find the cross-products and solve for the unknown number.

$$? \times 24 = 204 \times 1$$
$$? = \frac{204}{24} = 8.5$$

ANSWER: There are **8.5 days** in 204 hours.

Change each measurement to a mixed number.

28. 90 sec = _____ min

29. 2,190 hr = _____ days

30. 75 min = _____ hr

31. 36 hr = _____ days

32. 105 min = _____ hr

33. 80 sec = _____ min

34. 150 sec = _____ min

35. 50 hr = _____ days

36. 100 min = _____ hr

Solve each problem.

37. Jerry is in school for 8 hours each week day. How many minutes does he spend in school each week?

38. Tom, a loan officer at a bank, gives a 36-month car loan to a customer. The customer wants to know how many years it will take to pay off the loan. What should Tom say?

39. Jim works 8 hours a day, 5 days a week. The payroll clerk needs to know how many hours Jim works in one week so she can process Jim's time card. How many hours does Jim work in one week?

40. How many hours of work does Jim miss when he takes a two-week vacation?

41. Sally packages carpentry nails at a factory. She can package 100 nails in a box every 15 seconds. How many minutes will it take her to package 200 nails?

42. Dario, a charter bus driver, must record the number of hours he drives. One week he made 12 trips of 40 minutes each between a hotel and an airport. How many hours should he record for those trips?

43. Michelle spends 3 hours per week in an English 101 class. How many minutes is she in class each week?

44. Mike is an assistant in a college science department. One of his duties is to tutor for 150 minutes each week. If he runs a one-hour tutoring session twice a week for a biology class, how many more minutes does he need to tutor during the week?

45. Taneisha is a baker. The baking time for a banana bread recipe is 90 minutes. She needs to set the oven timer in hours and minutes. For how many hours and minutes should she set the timer?

46. Jean watched a basketball game on television. Her team took four time outs, each one 45 seconds long. How much total time did the team take in time outs in minutes?

47. Cheri is a physician's assistant. She needs to order thermometers for taking patients' temperatures. Brand A gives a reading in 2 minutes. Brand B gives a reading in 100 seconds. Which brand is faster? How much faster is it?

Adding and Subtracting Measurements of Time

When you add or subtract time measurements, you may have to carry or borrow between the different units. Here are two examples.

EXAMPLE 1 Barb worked 6 hours and 50 minutes on Tuesday and 5 hours 30 minutes on Thursday. How much total time did she work on the two days?

STEP 1 Line up the measurements, putting like units under like units.

$$\begin{array}{r} 6 \text{ hr } 50\text{min} \\ + 5 \text{ hr } 30 \text{ min} \\ \hline \end{array}$$

STEP 2 Add the minutes and add the hours.

$$\begin{array}{r} 6 \text{ hr } 50\text{min} \\ + 5 \text{ hr } 30 \text{ min} \\ \hline 11 \text{ hr } 80 \text{ min} \end{array}$$

STEP 3 80 minutes is 1 hour and 20 minutes. Rewrite the sum.

$$11 \text{ hr} + 80 \text{ min}$$
$$= 11\text{hr} + 1 \text{ hr} + 20 \text{ min}$$
$$= 12 \text{ hr} + 20 \text{ min}$$

ANSWER: Barb worked **12 hours and 20 minutes** on the two days.

Add.

1. 2 years   7 months
  + 8 years 10 months

2. 8 weeks 4 days
  + 10 weeks 6 days

3. 10 min 43 sec
  + 12 min 38 sec

EXAMPLE 2 Subtract 6 minutes 45 seconds from 12 minutes 20 seconds.

STEP 1 Line up the measurements, putting like units under like units.

$$\begin{array}{r} 12 \text{ min } 20 \text{ sec} \\ - 6 \text{ min } 45 \text{ sec} \\ \hline \end{array}$$

STEP 2 One minute is the same as 60 seconds. Rewrite the top number as 11 minutes + 60 seconds + 20 seconds, or 11 minutes 80 seconds.

$$\begin{array}{r} 11 \text{ min } 80 \text{ sec} \\ - 6 \text{ min } 45 \text{ sec} \\ \hline \end{array}$$

STEP 3 Subtract the seconds and subtract the minutes.

$$\begin{array}{r} 11 \text{ min } 80 \text{ sec} \\ - 6 \text{ min } 45 \text{ sec} \\ \hline 5 \text{ min } 35 \text{ sec} \end{array}$$

ANSWER: The result of the subtraction is **5 minutes 35 seconds**.

Subtract.

4. 6 hours 40 min
  − 3 hours 15 min

5. 7 days   6 hours
  − 2 days 12 hours

6. 12 years 18 weeks
  − 7 years 10 weeks

Add or subtract.

7.   17 hr   0 min
      − 6 hr 45 min

8.   32 hr 7 min
      − 18 hr

9.   6 days   7 hr
      − 2 days 20 hr

10.   6 hr
       + 3 hr 25 min

11.   16 days 10 hr
       + 12 days   8 hr

12.   16 min 48 sec
       + 32 min 45 sec

13.   28 min 30 sec
       − 16 min 48 sec

14.   3 mo
       − 2 mo 1 wk

15.   21 days 17 hr
       + 6 days 18 hr

Solve each problem.

16. Sol is a human resources assistant. He keeps track of how much vacation time each employee earns. An employee earned 4 hours 30 minutes vacation time last week and 6 hours 45 minutes vacation time this week. How much earned vacation time should Sol record for the employee?

17. Jay is a benefits and retirement coordinator. One of his responsibilities is to keep track of employees' time employed. An employee began working 18 years ago, but took a leave of absence during that time for 1 year 3 months. The leave of absence time does not count as time worked when calculating retirement eligibility. How much time should Jay count for that employee's retirement eligibility?

18. Ronette is a telemarketer. A computer keeps a record of how long she is on the phone for each call she makes. Her supervisor wants to know the total amount of time she was on the phone for her last three calls. Her times were 3 minutes 28 seconds, 2 minutes 47 seconds, and 5 minutes 55 seconds. What was her total?

19. Kate is a management consultant studying the efficiency of a company's workforce. She asks employees to record how much time they spend on actual work tasks every day for a week. One employee recorded these times: 6 hours 48 minutes; 7 hours 21 minutes; 5 hours 57 minutes; 7 hours 46 minutes; and 6 hours 38 minutes. What total should Kate record for that employee?

20. Jackie has a 6-hour telephone card. She has used 2 hours 39 minutes. How much time does she have left on the card?

Multiplying and Dividing Measurements of Time

For some problems you can multiply or divide a time measurement by a number. For example, if you know the total time someone worked during a week, you can divide to find out the average time worked each day. If your commute to school or work takes the same number of minutes each day, you can multiply to find out how much time you spend commuting each month.

EXAMPLE 1 A survey researcher asked the employees at a supermarket how many hours they worked last week. One employee answered 42 hours in 5 days. What was the average length of that employee's workday, in hours and minutes?

STEP 1 Decide which operation to use.   *The operation is division.*

STEP 2 Write the problem and divide. $\frac{42}{5} = 8\frac{2}{5}$ *hours*

STEP 3 To change $\frac{2}{5}$ hour to minutes, multiply $\frac{2}{5}$ times the number of minutes in an hour. $\frac{2}{5} \times 60 = 24$

ANSWER: The employee's average workday was **8 hours 24 minutes**.

Divide.

1. 6 hours 10 minutes ÷ 5

2. 13 weeks 5 days ÷ 6

3. 8 months 20 days ÷ 4
   (Use 1 month = 30 days.)

4. 46 hours 20 minutes ÷ 5

EXAMPLE 2 Multiply 6 weeks 3 days by 5.

STEP 1 Write the problem.

$$\begin{array}{r} 6 \text{ weeks } 3 \text{ days} \\ \times \qquad\quad 5 \\ \hline \end{array}$$

STEP 2 Multiply the number of days and the number of weeks by 5.

$$\begin{array}{r} 6 \text{ weeks } 3 \text{ days} \\ \times \qquad\quad 5 \\ \hline 30 \text{ weeks } 15 \text{ days} \end{array}$$

STEP 3 15 days is the same as 2 weeks and 1 day.

30 weeks plus 15 days
= 30 weeks plus 2 weeks
plus 1 day
= 32 weeks 1 day

ANSWER: The product is **32 weeks 1 day**.

Multiply or divide.

5. 2 years 4 months × 4

10. 6 months 3 days × 3

6. 15 min 20 sec × 5

11. 6 days 14 hours × 5

7. 3 years 8 months ÷ 2

12. 12 days 18 hours ÷ 3

8. 3 days 8 hours ÷ 4

13. 6 min 38 sec × 5

9. 16 years 4 months × 3

14. 7 hours 12 minutes × 4

Solve each problem.

15. Thomas is a factory supervisor. He sets the speed on an assembly line so that the assembly of a bicycle is completed every 5 minutes 20 seconds. To help plan breaks, Thomas calculates times required to complete different numbers of bicycles. How long will it take to complete 25 bicycles?

16. Kym is a nurse's aid in a nursing home. To complete a work report, she needs to record the average amount of time she spends per resident giving medications. If she spent 48 minutes 24 seconds giving medications to 8 residents, how much time did she average with each resident?

17. During the past 10 years 3 months, Ryan has had three different jobs. What is his average time for each job?

18. Emily is an aerobics instructor at a fitness center. She teaches a special class in which the students exercise 45 minutes each day, including weekends. How many hours does a student in that class exercise during a five-week period?

19. The water pump at the Jemson's house runs about $2\frac{1}{2}$ hours daily. How many hours does the pump run weekly?

Comparing and Ordering Time Measurements

In order to compare the measurements of two time intervals, it is easier if the measurements use the same unit.

EXAMPLE 1   The directions on a box of cake mix say to beat the batter for 2 minutes. Joel watched the second hand on the kitchen clock and stopped mixing after 90 seconds. Did he beat the cake mix batter long enough?

STEP 1   Write the two measurements.   *90 sec, 2 min*

STEP 2   Rewrite one of the measurements so they have   *90 sec, 120 sec*
the same unit. This example uses seconds.

STEP 3   Compare the measurements.   *90 sec < 120 sec*

ANSWER: Since 90 seconds is less than 2 minutes, Joel did not beat the cake mix batter long enough.

EXAMPLE 2   Which is shorter, 3 weeks or 18 days?

STEP 1   Write the two measurements.   *3 weeks, 18 days*

STEP 2   Rewrite one of the measurements so they have   *21 days, 18 days*
the same unit. This example uses days.

STEP 3   Compare the measurements.   *21 days > 18 days*

ANSWER: The shorter period of time is **18 days**.

Circle the longer period of time.

1. 3 days or 80 hours

2. 360 minutes or 4 hours

3. 10 weeks or 3 months

4. 2 years or 22 months

5. 3 hours or 30,000 seconds

6. 72 hours or $3\frac{1}{2}$ days

7. 8 months or 274 days

8. 6 weeks or 50 days

9. 3 years or 160 weeks

10. 248 minutes or 4 hours

11. 6 weeks or 2 months

12. $3\frac{1}{2}$ years or 65 weeks

Fill in each blank with <, >, or =.

13. 2 days _____ 48 hours

14. 3 weeks _____ 24 days

15. 6 years _____ 56 months

16. 18 hours _____ 475 minutes

17. 70 months _____ 26 weeks

18. $6\frac{1}{2}$ hours _____ 420 minutes

19. 200 minutes _____ $3\frac{1}{2}$ hours

20. 17 weeks _____ 4 months

21. $3\frac{1}{2}$ years _____ 42 months

22. 180 seconds _____ 5 minutes

23. 30 hours _____ 2 days

24. 4 days _____ 96 hours

Write the measurements from shortest to longest.

25. 3 minutes, 200 seconds, 45 seconds

_____

26. $2\frac{1}{2}$ days, 65 hours, 420 minutes

_____

27. 36 weeks, 10 months, 1 year

_____

28. $2\frac{1}{2}$ years, 40 months, 85 weeks

_____

Solve each problem.

29. Mary has 13 vacation days. Jose has 3 weeks. Who has more vacation days?

30. A firefighter was on duty for 36 hours non-stop. Was that more or less than 2 days?

31. Manuel volunteered as an aide at the nursing home 40 hours last month. Is his volunteer time more or less than 3,000 minutes?

32. Sam, a manager of a hardware and home supply store, thinks it should take 15 minutes to assemble a barbeque grill for a customer. An employee tells Sam she completed the job in a quarter hour. Is the amount of time the employee spent more than, less than, or the same as what Sam thinks is adequate?

33. The owner of a landscaping business asked two of his supervisors how much time they spent on sales calls last week. Joe said 6 hours and Masao said 400 minutes. Which supervisor reported more time on sales calls?

Focus on Algebra: Velocity

When an object or person is moving, we can measure its **velocity.** Velocity is the same as speed or rate. When an object is traveling at a constant rate or velocity, we can use a formula to calculate that velocity. If an object travels a distance d in a time t, then the average velocity r can be calculated from the formula $r = \frac{d}{t}$.

EXAMPLE 1 On vacation with her family, Maria drove 180 miles in 3 hours. What was the average velocity for the trip?

STEP 1 Write the formula. $r = \frac{d}{t}$

STEP 2 Substitute numbers from the problem into the formula. $r = \frac{180}{3}$

STEP 3 Divide to simplify the fraction. $r = 60$

ANSWER: The rate is **60 miles per hour (mph)**.

Use the formula $r = \frac{d}{t}$ to find each missing velocity.

1. $r =$ _____
    $d = 7.5$ mi
    $t = 3$ hr

2. $r =$ _____
    $d = 1{,}800$ mi
    $t = 6$ hr

3. $r =$ _____
    $d = 180$ mi
    $t = 1.5$ hr

You can use the formula $r = \frac{d}{t}$ to find the distance if you know the rate and time.

EXAMPLE 2 Jon, a bus driver, drives for the first 3 hours of a trip at 55 mph. A passenger asks how many miles they have traveled. What should be Jon's answer?

STEP 1 Write the distance formula. $r = \frac{d}{t}$

STEP 2 Substitute numbers from the problem into the formula. Be sure to substitute each number for the correct **variable**, or letter. $55 = \frac{d}{3}$

STEP 3 Multiply both sides of the equation in Step 2 by 3. $165 = d$

ANSWER: Jon should answer that they have traveled **165 miles**.

Use the formula $r = \frac{d}{t}$ to find each missing distance.

4. $r = 400$ mph
    $d =$ _____
    $t = 22.5$ hr

5. $r = 10$ mph
    $d =$ _____
    $t = 6\frac{1}{2}$ hr

6. $r = 60$ mph
    $d =$ _____
    $t = 2\frac{3}{4}$ hr

There are times when you might want to know how long a trip will take at a particular average velocity. For example, a truck driver or moving van company might need to know how long a 414 mile trip would take at an average velocity of 52 miles per hour.

EXAMPLE 3   Jarod is a truck driver. He is beginning a 372-mile trip to deliver an order to a customer. He needs to tell the customer an estimated arrival time. From experience, he knows his average velocity will be about 62 mph. How long will the trip take if he makes no stops?

STEP 1 Write the formula. $r = \dfrac{d}{t}$

STEP 2 Substitute numbers from the
        problem into the formula. $62 = \dfrac{372}{t}$

STEP 3 Solve the equation. $62t = 372$

ANSWER: The trip will take about **6 hours**. $t = \dfrac{372}{62} = 6$

Use the formula $r = \dfrac{d}{t}$ for each exercise.

7. $r = 61$ mph   10. $r = 150$ mph   13. $r = 40$ mph
    $d = 366$ mi $d = 300$ mi $d = 220$ mi
    $t = $ _____ $t = $ _____ $t = $ _____

8. $r = $ _____   11. $r = 60$ mph   14. $r = 65$ mph
    $d = 1{,}050$ mi $d = 135$ mi $d = $ _____
    $t = 3\frac{1}{2}$ hr $t = $ _____ $t = 8$ hr

9. $r = 60$ mph   12. $r = $ _____   15. $r = 50$ mph
    $d = $ _____ $d = 480$ mi $d = 600$ miles
    $t = 4\frac{1}{2}$ hr $t = 1\frac{1}{3}$ hr $t = $ _____

Solve each problem.

16. An airline pilot flew a 1,300-mile flight in $3\frac{1}{4}$ hours. To complete a report, he needs to record the average velocity. What was the average velocity for the flight?

17. Jason rode his bike $27\frac{1}{2}$ miles in 2 hours. What was his average velocity?

18. Jackie made a 150-mile trip in $2\frac{1}{2}$ hours. What was her average speed?

19. Fred is a branch manager for a car rental company. A customer reported that the cruise control feature on a rental car was not working properly. To test it, Fred used the car for a weekend trip. He drove 203 miles with the cruise control set so that the average speed was 58 mph. How many hours did the trip take if the cruise control was working properly?

Section 5 Cumulative Review

Rewrite each measurement so it uses the given unit.

1. _____ seconds = 1 minute

2. 1 year = _____ months

3. _____ hours = 1 day

4. 1 month ≈ _____ weeks

5. 1 week = _____ days

6. _____ minutes = 1 hour

7. _____ weeks = 1 year

8. 1 month ≈ _____ days

9. 3 days = _____ hr

10. _____ min = $2\frac{1}{2}$ hr

11. 4 wk = _____ days

12. $4\frac{1}{2}$ yr = _____ mo

13. 150 sec = _____ min

14. 28 mo = _____ yr

15. 147 days = _____ wk

16. _____ hr = 270 min

Circle the longest time period.

17. 390 minutes; 5 hours; 4,200 seconds

18. 2 years; 110 weeks; 30 months

19. 18 days; 2 weeks; 300 hours

20. 6 months; 30 weeks; 1,080 hours

21. $2\frac{1}{2}$ minutes; 100 seconds; $\frac{1}{2}$ hour

22. 13 months; 1 year; 60 weeks

Fill in each blank with >, <, or =.

23. 6 years _____ 72 months

24. $4\frac{1}{2}$ days _____ 90 hours

25. 8 weeks _____ 60 days

26. 330 minutes _____ $6\frac{1}{2}$ hours

Mark the given duration on each stop watch.

27. 1 min 10 sec

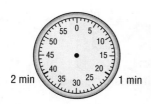

28. 2 min 42 sec

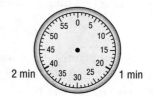

Add, subtract, multiply, or divide.

29. 6 hr 30 min − 2 hr 45 min

30. 3 days 18 hr × 5

31. 2 yr 7 mo + 3 yr 8 mo

32. 7 min 18 sec ÷ 6

33. 10 min 12 sec − 3 min 40 sec

34. 7 hr 45 min × 3

35. 6 wk 5 days + 3 wk 4 days

36. 5 days 3 hr 16 min ÷ 2

Circle the better estimate for each duration.

37. time to drink a 12-ounce soft drink   20 min or 20 sec

38. vacation time after 10 years on a job   15 wk or 15 days

39. time to microwave popcorn $2\frac{1}{2}$ sec or $2\frac{1}{2}$ min

40. time to sew a dress   17 hr or 17 days

41. time on a shift at a factory   8 days or 8 hr

Solve each problem.

42. George has 20 days of sick leave accumulated at his workplace. He must have surgery, and will be away from work for 6 weeks. Does he have enough sick days for his recovery period? (Assume he works a 5-day work week.)

43. Henri worked these hours during one week: 7 hours 50 minutes; 8 hours 20 minutes; 8 hours 10 minutes; 7 hours 45 minutes; and 8 hours 55 minutes. What was his average daily work time?

44. Carol exercises 1 hour 20 minutes daily. If she does this 6 days each week, how much time does she spend each week exercising?

45. Josie is the manager of a retail clothing store. Employees are supposed to have a 15-minute break every 3 hours. Luke has worked four 50-minute shifts at checkout registers with no break. Should Josie tell Luke to take a break?

Solve each of the following problems about time periods and health medications.

46. A nurse practitioner tells Joe to take a pill twice a day, spaced at equal time intervals. Joe asks how many hours apart he should take his pills. How should the nurse practitioner answer?

47. Joe took his first pill at 6:15 P.M. When should he take his second pill?

48. A dentist removes Ann's wisdom teeth and then tells her to take a pain killer every four hours. Ann takes her first pill at 2:30 P.M. She calls the dentist to ask what times she needs to take the next three pills. What times should the dentist tell her?

49. Mary is a nanny and is taking care of a child while the parents are on a trip. She wants to give the child two aspirin tablets for a fever. The aspirin bottle warns not to take more than 12 tablets in a 24-hour period. How often can Mary give the child a pair of aspirin during a 24-hour period?

50. If Mary gave the first two aspirin to the child at 7:00 A.M., at what times could she give another two aspirins over the next 24 hours?

51. Jess has received an antibiotic for his bronchitis. The instructions say to take four capsules, three times a day. If he takes his first set of four capsules at 6:45 A.M., at what time should he take his next set of capsules?

Find the volume of each container.

52.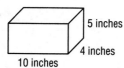
5 inches
4 inches
10 inches

53.
cube
1.5 meters

54.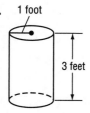
1 foot
3 feet

55.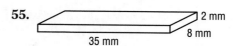
2 mm
8 mm
35 mm

56.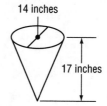
14 inches
17 inches

57.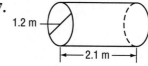
1.2 m
2.1 m

Find each indicated length.

58.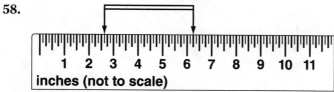
inches (not to scale)

59.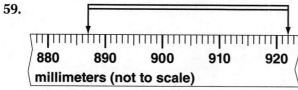
millimeters (not to scale)

Mark the given measurement on each set of dials.

60. 25,033
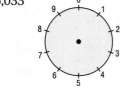

61. 40,995
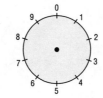 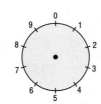 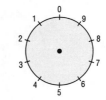 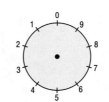 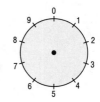 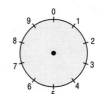

USING
NUMBER
POWER

Home Remodeling

Jeff and Marisa are remodeling their home. Here are some of the problems and questions they have encountered.

1. On the roof, Jeff wants to put shingles on a rectangle 25 feet by 20 feet. What is the area of that part of the roof? What is the perimeter of that part of the roof?

2. Jeff bought enough shingles to cover 575 square feet. How many square feet of shingles will be left over after he covers the 25-foot by 20-foot rectangle?

3. Marisa is installing new windows. Each window is 2 feet by 4 feet. She installed 120 square feet of windows. How many windows did she install?

4. Jeff and Marisa put siding on their house. The siding they bought comes in 12-foot lengths. The front of their house is 32 feet wide. If they use three strips of siding to go across the front of the house, how much extra is there?

5. The strips of siding cover an area that measures 12 feet by 6 inches. How many strips will they use to cover 600 square feet?

6. Marisa bought scraps of plywood to build a dog house. The pieces were 2 feet wide, and their lengths were $3\frac{1}{2}$ feet, $2\frac{1}{4}$ feet, and 6 feet. What was the total length of the scraps of plywood? (Your answer should be in feet and inches.)

7. Asphalt for covering the driveway comes in 5-gallon buckets. Each bucket can cover 100 square yards. Their driveway is 100 feet long and 18 feet wide. How many buckets will they need? (Hint: 9 sq ft = 1 sq yd)

8. The angle formed by the two parts of the roof is 150°. Draw an angle of 150°. Is that angle acute, right, or obtuse?

9. In the attic, the chimney forms an angle of 75° with the floor. Draw an angle of 75°. Then find the complement and the supplement of the angle.

10. On the stairs, the handrail forms an angle of 50° with the vertical posts. Draw an angle of 50°. Then draw and label the complement and the supplement of a 50° angle.

Here are "before" and "after" plans for Jeff and Marisa's remodeling project.

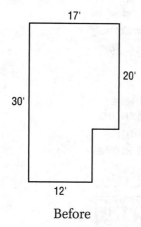

Before

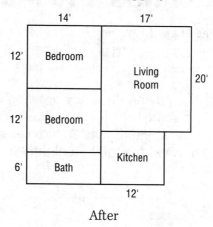

After

11. What is the perimeter of the house before the remodeling?

_____

12. What is the perimeter of the house after the remodeling?

_____

13. How many square feet were added to the house?

_____

14. Looking at the "after" picture, find the area of the living room, bathroom, and kitchen.

_____

15. Looking at the "after" picture, write the measurements for the living room and kitchen in yards.

Living Room _____

Kitchen _____

16. Looking at the "before" and "after" pictures of the remodeled house, what rooms were made in the new part of the house?

_____

Scale on Drawings and Maps

On a map or a scale drawing of an object, lengths in the drawing represent actual lengths. For example, if the scale for a drawing of a building is "1 inch = 10 feet," then 1 inch on the drawing represents 10 feet in the actual building.

EXAMPLE 1 Dave is a builder studying the scale drawing of a kitchen shown below. The scale is "1 inch = 4 feet." He has measured the drawing with a ruler and written his measurements in inches as shown. What are the actual dimensions of the stove?

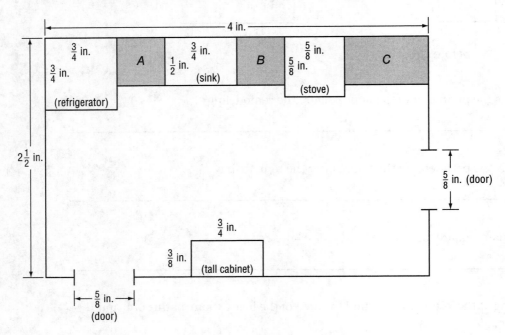

STEP 1 Write the scale as a ratio.

$$\frac{1\ inch}{4\ feet}$$

STEP 2 Set up a proportion using two ratios. Use "?" to represent the unknown distance.

$$\frac{1}{4} = \frac{\frac{5}{8}}{?}$$

STEP 3 Cross multiply and solve for the unknown distance.

$$1 \times ? = 4 \times \frac{5}{8}$$

$$? = \frac{\overset{1}{\cancel{4}}}{1} \times \frac{5}{\underset{2}{\cancel{8}}} = \frac{5}{2} = 2\frac{1}{2}$$

ANSWER: The stove is **$2\frac{1}{2}$ feet by $2\frac{1}{2}$ feet**, or 2 feet 6 inches by 2 feet 6 inches.

Solve each problem.

1. What are the actual dimensions of the refrigerator?

2. What are the actual dimensions of the tall cabinet?

3. What are the actual overall dimensions of the kitchen?

4. The shaded regions A, B, and C are countertops. What is the actual combined overall length of the three countertops?

EXAMPLE 2 Jana is looking at a map with a scale of "2 inches = 150 miles." On the map, the distance between Chicago and her hometown is 3 inches. What is the distance between her hometown and Chicago?

STEP 1 Write the scale as a ratio.

$$\frac{2 \text{ inches}}{150 \text{ miles}}$$

STEP 2 Set up a proportion using two ratios. Use "?" to represent the unknown distance.

$$\frac{2}{150} = \frac{3}{?}$$

STEP 3 Cross multiply and solve for the unknown distance.

$$2 \times ? = 3 \times 150$$
$$2 \times ? = 450$$
$$? = 225$$

ANSWER: Jana's hometown is about **225 miles** from Chicago.

Use each scale and measurement to find the actual length.

5. Scale: 1 in. = 6 ft
Measurement: 6 in.
Actual size: _____

6. Scale: 3 in. = 25 miles
Measurement: 12 in.
Actual size: _____

7. Scale: 2 in. = 75 ft
Measurement: 10 in.
Actual size: _____

8. Scale: 2 cm = 70 m
Measurement: 15 cm
Actual size: _____

9. Scale: 1 in. = 30 ft
Measurement: $4\frac{1}{2}$ in.
Actual size: _____

10. Scale: 1 cm = 10 km
Measurement: 12 cm
Actual size: _____

Use the scale "1 inch = 30 miles" for each problem.

11. On a map, the distance between Omaha and Des Moines is $4\frac{1}{2}$ inches. How far apart are Omaha and Des Moines?

12. The distance on a map between Kansas City and Omaha is 6 inches. How many miles is that?

Windchill

When the wind is blowing, a cold day feels even colder! This effect is called windchill. The following chart gives you the windchill temperature, or what the temperature feels like, for different wind speeds and different actual temperatures.

Windchill Table					
Actual Temperature (°F)	**40°**	**30°**	**20°**	**10°**	**0°**
Wind speed (mph)					
5	36°	25°	13°	1°	−11°
10	34°	21°	9°	−4°	−16°
15	32°	19°	6°	−7°	−19°
20	30°	17°	4°	−9°	−22°
25	29°	16°	3°	−11°	−24°
30	28°	15°	1°	−12°	−26°
35	28°	14°	0°	−14°	−27°
40	27°	13°	−1°	−15°	−29°

EXAMPLE 1   What is the windchill when the temperature is 10° and the wind speed is 15 mph?

STEP 1   Locate "15 mph" along the left and "10°" along the top.

STEP 2   Find the table entry where the "15 mph" row meets the "10°" column.

ANSWER: The windchill is −7°F. When the actual temperature is 10° and the wind is blowing at 15 miles per hour, the temperature feels like it is 7 degrees below zero on a calm day.

Use the windchill table to answer each question.

1. Are the temperatures given in °C or °F?

2. What is the lowest actual temperature listed on the chart?

3. If the temperature is 30°F and the wind speed is 15 miles per hour, what is the windchill?

EXAMPLE 2 What combination of actual temperature and wind speed result in a windchill of 16°?

STEP 1 Locate "16°" in the table.

STEP 2 From the entry of "16°," read the actual temperature and wind speed for that entry.

ANSWER: According to the windchill table, the conditions that give you a windchill of 16° are:
actual temperature of 30°F, wind speed of 25 mph

Use the windchill table to answer each question.

4. If the windchill is −4°F and the actual temperature is 10°F, what is the wind speed?

5. If the windchill is −27°F and the wind speed is 35 mph, what is the actual temperature?

6. At what wind speeds is the windchill colder than −26°F?

7. If the temperature is 20°F and the wind speed is 20 miles per hour, what is the windchill?

8. If the temperature is 10°F and the wind speed is 30 miles per hour, what is the windchill?

9. If the temperature is 0°F and the wind speed is 30 miles per hour, what is the windchill?

10. At what conditions is the windchill 19°F?

11. At what conditions is the windchill −24°F?

Doctors' Offices

1. Lucy weighed 124 pounds on her last doctor's visit. The nurse weighed Lucy today and said she had gained 12 pounds. How much does Lucy weigh? Draw a weight scale and show that weight on the scale.

2. When the doctor saw Lucy's weight gain, he said she needed to lose 8 pounds. How much will she weigh after she loses the weight?

3. Jerry, a veterinarian, wanted to weigh his dog Tugger. Jerry weighed himself. The scale read 196 pounds. Jerry held Tugger and weighed himself again. The scale read 225 pounds. How much does Tugger weigh?

4. While Karma was pregnant, she gained weight every month. The nurse recorded her weight gain each month.

Jan.	2 lb 5 oz	Apr.	3 lb 1 oz	July	6 lb 3 oz
Feb.	2 lb 4 oz	May	3 lb 7 oz	Aug.	4 lb 10 oz
Mar.	1 lb 12 oz	June	2 lb 5 oz	Sept.	5 lb 15 oz

   a. What was her total weight gain? Record your answer in pounds and ounces.

   b. What was Karma's average weight gain over the 9-month period?

5. Karma had twin girls. When they were born, their total weight was 16 lb 6 oz. Their total weight has tripled since their birth. How much do they weigh now?

6. What is the average weight of each twin now?

7. What was the average weight of each twin when they were born? Draw a weight scale and show that weight on the scale.

8. Jose was not feeling well. The nurse took his temperature and found out he had a fever of 102.5°F. Draw a thermometer and show that temperature on the thermometer. How much above 98.6°F was his temperature?

9. During an operation, doctors lowered Mai's body temperature. At the beginning of the operation, her body temperature was 94.9°F. Draw a thermometer and show that temperature on the thermometer. How much below 98.6°F was her temperature?

10. Tyler's temperature was 99.2°F at 10 A.M. and 101.9°F at 3 P.M. Draw a thermometer and show those temperatures on the thermometer. By how much did his temperature rise?

11. The Jones' baby was sick. When the nurse took his temperature, it was 104.2°F. Draw a thermometer and show that temperature on the thermometer. How much above 98.6°F was the baby's temperature?

12. Joyce's temperature was 102.4°F, so she took some medicine. Later it was 99.9°F. Draw a thermometer and show those temperatures on the thermometer. By how much did her temperature drop?

13. During a routine physical, Fred's temperature was normal at 97.4°F. Draw a thermometer and show that temperature on the thermometer. By how much does Fred's normal temperature differ from 98.6°F?

Automotive Fluids

The tables on this page show approximate amounts of fluids recommended for several cars by an automotive fluid manufacturer. The cars are identified by engine codes.

	Fluid Amounts					
	Coolant		**Automatic Transmission Fluid**		**Manual Transmission Fluid**	
Engine Code	**Liters**	**Quarts**	**Liters**	**Quarts**	**Liters**	**Pints**
[G]	7.6	8	8.2	$8\frac{5}{8}$	2.8	6
[5]	7.6	8	8.2	$8\frac{5}{8}$	3.0	$6\frac{1}{4}$
[N]	12.5	$13\frac{1}{4}$	11.2	$11\frac{7}{8}$	2.6	$5\frac{1}{2}$
[H]	13.0	$13\frac{3}{4}$	11.2	$11\frac{7}{8}$	3.1	$6\frac{1}{2}$

1. How many liters of coolant (antifreeze) does the cooling system of a car with engine [N] hold?

2. For a car with engine [G], how many more liters of transmission fluid are needed for an automatic transmission than for a manual transmission?

3. For a car with engine [N], how many more *pints* of transmission fluid are needed for an automatic transmission than for a manual transmission?

4. If a mechanic fills the cooling systems for a car with engine [N] and a car with engine [H], how many liters of coolant does he use in all?

5. How many more liters of coolant are needed to fill the cooling system of a car with engine [N] than a car with engine [5]?

6. A mechanic replaces the automatic transmission fluid in 2 cars with engine [G], a car with engine [5], and 3 cars with engine [H]. How many quarts of fluid does she use in all?

7. A mechanic starts with a full 20-gallon container of coolant. He fills the cooling systems of a car with engine [G] and 3 cars with engine [N]. How much coolant is left in the container?

8. Change each of the oil amounts below from quarts to pints.

Amount of Oil Needed for an Oil Change (replacing filter)		
Engine Code	**Quarts**	**Pints**
[G]	$4\frac{5}{8}$	___?___
[5]	$5\frac{1}{4}$	___?___
[N]	5	___?___
[H]	$6\frac{1}{8}$	___?___

Recipes

1. Cheri is having a bridal shower for a friend. She wants to make strawberry punch for 25 people. At the store, the lemonade cans contain 12 ounces of frozen lemonade. How many cans should she buy?

2. How many ounces of frozen lemonade will Cheri have left over after she prepares the punch?

3. The strawberry punch recipe calls for seven 16-ounce cans of water. How many cups of water is this?

4. Margaret plans to make baked chocolate pudding for the bridal shower. The recipe calls for $\frac{1}{4}$ cup shortening and 1 cup flour. How many times will she need to fill a quarter-cup measuring container to measure out the flour? How many times will she fill the quarter-cup container to measure out the shortening?

5. The baked chocolate pudding recipe calls for 3 tablespoons of cocoa. Using a half-teaspoon measuring spoon, how many times will she need to fill it to measure out 3 tablespoons?

6. Margaret is going to double the recipe for baked chocolate pudding. Write a list of the ingredients and their amounts that Margaret will need.

7. For another party, Cheri is going to make strawberry punch using half of the recipe. Write a list of ingredients and their amounts for half the strawberry punch recipe.

Strawberry Punch

one 16-oz can frozen orange juice

one 16-oz can frozen lemonade

one 16-oz pkg. frozen strawberries

one 32-oz bottle ginger ale

seven 16-oz cans water

Mix all ingredients in punch bowl and top with fruit slices and ice rings. Serves approximately 25.

Baked Chocolate Pudding

$\frac{1}{4}$ c shortening	Combine shortening, salt, cinnamon and sugar. Cream thoroughly. Sift flour, baking powder, baking soda, and cocoa together. Add to creamed mixture alternately with milk, blending well after each addition. Add nuts. Pour batter into greased pan. Bake at 350° for 45 minutes.
1 c sifted flour	
$\frac{1}{2}$ tsp salt	
2 tsp baking powder	
1 tsp cinnamon	
$\frac{1}{2}$ tsp baking soda	
$\frac{1}{4}$ c sugar	
$\frac{2}{3}$ c milk	
3 tbsp cocoa	
$\frac{1}{2}$ c chopped nuts	

Ice, Water, Gas

1. Jeff has a job with a company that delivers ice. He delivers 25 pounds of ice cubes to various restaurants in town. Each side of each cube is 300 millimeters. What is the volume of one cube of ice?

2. Jeff also delivers bags of ice for ice machines in convenience stores. Each bag is $1\frac{1}{2}$ feet long, $\frac{1}{2}$ foot wide, and 8 inches across. Find the volume of a bag of ice.

3. On a winter day, water dripped from a branch of a tree and formed a cone of ice. The height of the cone is 18 inches and the radius of the base of the cone is $\frac{1}{2}$ foot. What is the volume of the ice cone?

4. Gabrielle has a new bathtub. The bathtub is 60 inches long, 31 inches wide, and 20 inches high. If she fills the tub full, what is the volume of water in the tub?

5. When Gabrielle bathes her baby, she fills the bathtub $\frac{1}{4}$ full. What is the volume of the water in the tub when she bathes her baby?

6. As a special treat for his birthday, Marco was given a chocolate ice cream cone. The ice cream parlor uses cones that have a height of 5 inches and a base with diameter of 2 inches. Find the volume of the cone.

7. As part of Marco's birthday celebration, he and his friends were driven to a field where a hot air balloon waited. The part of the balloon which carried the passengers, called a **gondola**, was a box-shaped basket that was 6 feet long, 5 feet wide, and 4 feet tall. Find the volume of the gondola.

8. When the hot air balloon is filled, it takes the shape of a ball. That shape is called a **sphere**. You can use the formula $V = \frac{4}{3}\pi r^3$ to find the volume of a sphere, where $\pi \approx 3.14$, r is the radius of the balloon, and r^3 means $r \times r \times r$. Find the volume of the hot air balloon with a radius of 4.5 meters.

9. Theresa makes juice as part of her job preparing breakfast at a restaurant. She uses a frozen concentrate that comes in a can shaped like a cylinder. The can is eight inches tall and has a diameter of 5 inches. To make the juice, she adds three containers of water to the frozen concentrate. What is the volume of the water that she adds to the frozen concentrate?

10. Theresa had a strawberry malt for a treat. As she drank it, she practiced her math. She measured the malt container and found it to be a cylinder 7 inches tall with a radius of 2 inches. What is the volume of the cylinder?

Heating and Air Conditioning

In order to design a heating and air conditioning system for a building, you have to know the interior volume of the building. For any building, the interior volume can be calculated by multiplying the number of square feet of floor area times the height of the ceiling.

$$V = \text{floor area} \times \text{ceiling height}$$

1. A room with floor area of 500 square feet and an 8-foot-high ceiling has what volume?

2. An apartment with a floor area of 1,800 square feet and an 8-foot-high ceiling has what volume?

3. A house has a volume of 16,000 cubic feet and has a floor area of 2,000 square feet. What is the height of the ceiling?

4. An office has a volume of 30,000 cubic feet. If the ceiling is 10 feet high, what is the floor area of the office?

5. What is the volume of a rectangular room with dimensions 15 feet by 12 feet and walls with a height of 10 feet?

6. A farmhouse with 1,400 square feet of living space has 10-foot-high walls. What is the volume of the farmhouse?

7. A grocery store is 1,000 feet long, 800 feet wide, and has walls 12 feet high. What is the volume of the store?

8. A store in a mall is a cube with volume 8,000 cubic feet. What is the height of the ceiling?

Packing and Moving

Use these volume formulas to solve the following problems.

Cube	$V = s^3$	where s is the side of the cube
Rectangular box	$V = \ell \times w \times h$	where ℓ, w, and h are the three dimensions of the box
Cylinder	$V = Bh$ $V = \pi r^2 h$	where B is the area of each round base, h is the height, $\pi \approx 3.14$, and r is the radius of the base
Cone	$V = \frac{1}{3}\pi r^2 h$	where $\pi \approx 3.14$, r is the radius of the base, and h is the distance between the tip of the cone and the base

1. Jones Moving Company was sent to the Smith residence to pack and move their furniture. The Smith children have a large supply of building blocks. The movers counted 156 blocks. Each cube block is 3 inches high. What is the volume of each building block? What is the total volume of all 156 blocks?

2. The movers want to pack the blocks in a rectangular box 3 feet long, 2 feet wide, and 1 foot high. What is the volume of the packing box in cubic feet?

3. Will the blocks all fit into one packing box? (Hint: Write both measurements in cubic inches.)

4. The movers have to carefully pack all of the Smiths' drinking glasses. Each glass is 5 inches tall with a diameter of 3 inches. What is the volume of each drinking glass?

5. The Smiths' end tables are shaped like a drum. Each table is 2 feet high and has a radius of 18 inches. What is the volume of each end table?

6. The Smiths' refrigerator is $5\frac{1}{2}$ feet high, 3 feet wide, and 3 feet long. Find the volume of the refrigerator.

7. The Smiths' kitchen table has cone-shaped legs. The movers removed the four legs and wondered if all four would fit into a long rectangular packing box 2 feet by 2 feet by 4 feet. Each table leg is 3 feet long with a radius of 2 inches. Find the volume of each table leg and the volume of the packing box.

8. The Smiths need to get their house ready to sell. They ordered a pile of gravel for their driveway. When it was delivered, it formed a cone that was 5 feet high with a diameter of 6 feet. What is the volume of the gravel?

Time Zones

Have you ever called a relative on the East coast or West coast and been asked "Why are you calling at this hour? You woke me up." You look at your clock and it's only 9:00 P.M.

PACIFIC MOUNTAIN CENTRAL EASTERN

The different regions of the country have different **time zones.** The different time zones mean that the clocks in our homes, schools, and businesses are coordinated with the apparent position of the sun.

Most of the United States is covered by four regions: Pacific, Mountain, Central, and Eastern. On the map, you can see the extent of each region. As you change regions from west to east, the time on the clock is moved ahead by one hour. For example, at 9 A.M. in the Pacific region, it is 10 A.M. in the Mountain region, 11 A.M. in the Central region, and noon in the Eastern region.

Use the time zone map to determine each of the following times.

1. 8 A.M. Pacific = _____ Central

2. Noon Mountain = _____ Pacific

3. 3 P.M. Eastern = _____ Mountain

4. 5:30 P.M. Central = _____ Eastern

5. 7 P.M. Pacific = _____ Eastern

6. 6:45 A.M. Eastern = _____ Pacific

7. 1:15 P.M. Mountain = _____ Pacific

8. 11 A.M. Mountain = _____ Pacific

9. 10:40 P.M. Central = _____ Mountain

10. Midnight Eastern = _____ Mountain

Solve each problem.

11. George lives in New York. He needs to talk to his business partner at 7:30 A.M., California time. What time should he call from New York?

12. Maria lives in Texas. At 9:00 P.M. she called her mother in Utah. What time was it at her mother's home?

13. Julia flew from Albuquerque, NM, to San Francisco, CA, an hour's flight. She left Albuquerque at 10:15 A.M. What time did she arrive in San Francisco?

14. Sam lives in Florida. He called his aunt in Seattle, which is on Pacific time. If he called at 5 P.M. Florida time, what time was it in Seattle?

15. Antoine needs to fly to Atlanta, GA, from Chicago, IL. If he leaves Chicago at 6 A.M., what time is that in Atlanta?

Wind Velocity

The velocity or speed of wind is measured by an anemometer. In 1805, Sir Francis Beaufort developed a wind scale to measure the effect wind had on a ship's sails. Wind is measured 10 meters (about 40 feet) above the ground.

		Beaufort Wind Scale	
Beaufort Number	Name	Miles per hour	Effect on land
0	Calm	less than 1	Calm; smoke rises vertically.
1	Light Air	1–3	Weather vanes inactive; smoke drifts with air.
2	Light Breeze	4–7	Weather vanes active; wind felt on face; leaves rustle.
3	Gentle Breeze	8–12	Leaves and small twigs move; light flags extend.
4	Moderate Breeze	13–18	Small branches sway; dust and loose paper blow about.
5	Fresh Breeze	19–24	Small trees sway; waves break on inland waters.
6	Strong Breeze	25–31	Large branches sway; umbrellas difficult to use.
7	Moderate Gale	32–38	Whole trees sway; difficult to walk against wind.
8	Fresh Gale	39–46	Twigs break off trees; walking against wind very difficult.
9	Strong Gale	47–54	Slight damage to buildings; shingles blown off roof.
10	Whole Gale	55–63	Trees uprooted; considerable damage to buildings.
11	Storm	64–73	Widespread damage; very rare occurrence.
12–17	Hurricane	74 and above	Violent destruction.

Find the Beaufort number for each exercise.

1. wind at 21 mph

2. light air

3. wind at 70 mph

Find the range of miles per hour for each exercise.

4. a strong breeze

5. Beaufort number 7

6. a gentle breeze

Describe the effect on land for each exercise.

7. wind of 49 mph

10. Beaufort number 10

8. fresh breeze

11. wind of 72 mph

9. calm air

12. Beaufort number 4

Miles Per Gallon

How far can your car go on a gallon of gasoline? You can calculate the **miles per gallon** (mpg) if you know the number of miles you drove and the number of gallons of gasoline you used.

EXAMPLE 1 Rollie put 11.6 gallons of gas in his car. He knew he had driven 290 miles since his last fill-up. How many miles per gallon (mpg) did he average?

STEP 1 Set up a proportion using the ratio $\frac{\text{miles}}{\text{gallons}}$.

$$\frac{290}{11.6} = \frac{?}{1}$$

STEP 2 Write the cross-products for the proportion.

$$? \times 11.6 = 290 \times 1$$

STEP 3 Divide each side by 11.6.

$$? = \frac{290}{11.6} = 25$$

ANSWER: Rollie got **25 miles per gallon** on the trip.

Find the number of miles per gallon for each exercise. (If necessary, round your answer to the nearest tenth.)

1. 280 miles driven
   14 gallons of gas
   mpg = _____

2. 486 miles driven
   16 gallons of gas
   mpg = _____

3. 114 miles driven
   6 gallons of gas
   mpg = _____

4. 142.8 miles driven
   6.8 gallons of gas
   mpg = _____

5. 375 miles driven
   12.5 gallons of gas
   mpg = _____

6. 1,700 miles driven
   70 gallons of gas
   mpg = _____

If you know how many miles per gallon a car gets, you can relate the number of miles you can drive to the number of gallons of gasoline you will need.

EXAMPLE 2 Juan's car usually gets about 32 mpg. If he has 7 gallons of gas in his car, how far can he expect to drive before he needs more gasoline?

STEP 1 Set up a proportion using the ratio $\frac{\text{miles per gallon}}{\text{gallons}}$.

$$\frac{32}{1} = \frac{?}{7}$$

STEP 2 Write the cross-products.

$$? \times 1 = 32 \times 7$$

STEP 3 Multiply.

$$? = 224$$

ANSWER: Juan can expect to drive **224 miles** before he needs more gasoline.

Find the distance that can be driven on the given amount of gas and the known miles per gallon. (If necessary, round your answer to the nearest tenth.)

7. has 8 gal of gas
   24 mpg
   distance = _____

9. has 14.5 gal of gas
   28.4 mpg
   distance = _____

11. has 40 gal of gas
    19.5 mpg
    distance = _____

8. has 6.7 gal of gas
   31 mpg
   distance = _____

10. has 6.2 gal of gas
    25.8 mpg
    distance = _____

12. has 27 gal of gas
    30 mpg
    distance = _____

Solve each problem.

13. Chu's car averages 21 mpg. He has 8.6 gallons of gas in his car. How far can he expect to drive before he needs more gasoline?

14. Dina's truck started a trip with a full tank of gas. She drove 1,800 miles, and then needed 90 gallons to refill her gas tank. What is the mpg for the truck?

15. For one part of a trip, Sara started with an odometer reading of 24,586 miles. On her first stop to fill the tank with gas, the odometer showed 24,886 miles. She needed 10 gallons of gas to fill the tank. Figure out the mpg for this part of her trip. (Hint: Subtract the odometer readings to get the mileage.)

16. Rod's car gets 27 mpg. If he used 46 gallons of gas during a three-day trip, how many miles did he travel?

17. Yuko put 14 gallons of gas in her car on Monday. If she usually gets 31 mpg, how far can she drive before she needs more gas?

18. Ryan's mother lives 250 miles from his home. He has 12 gallons of gas in his car. The car usually gets 26 mpg. Does he have enough gas to drive to his mother's home and return?

19. Samuel delivers supplies to medical centers. The longest drive he has is 220 miles each way. If his delivery van holds 18 gallons of gas and averages 23 mpg, will he have to buy gas on the trip?

20. Jackie's sports car averages 24 mpg. How far can she expect to drive on 12 gallons of gas?

Reading a Bus Schedule

Use this bus schedule to answer the following questions.

Eastbound							Westbound						
Crossroads	99-Maple	90-Maple	61-Maple	40-Cuming	23-Burt	16-Jackson	16-Jackson	23-Burt	40-Cuming	61-Maple	90-Maple	99-Maple	Crossroads
7:16	—	7:26	7:35	7:45	7:50	7:58	6:32	6:40	6:45	6:55	7:04	—	7:14
8:01	—	8:11	8:20	8:30	8:35	8:43	7:17	7:25	7:30	7:40	7:49	—	7:59
8:46	—	8:56	9:05	9:15	9:20	9:28	8:02	8:10	8:15	8:25	8:34	—	8:44
9:31	—	9:41	9:50	10:00	10:05	10:13	8:47	8:55	9:00	9:10	9:19	—	9:29
10:16	—	10:26	10:35	10:45	10:50	10:58	9:32	9:40	9:45	9:55	10:04	—	10:14
11:01	—	11:11	11:20	11:30	11:35	11:43	10:17	10:25	10:30	10:40	10:49	—	10:59
11:46	—	11:56	12:05	12:15	12:20	12:28	11:02	11:10	11:15	11:25	11:34	—	11:44
							11:47	11:55	12:00	12:10	12:19	—	12:29
12:31	—	12:41	12:50	1:00	1:05	1:13							
1:16	—	1:26	1:35	1:45	1:50	1:58	12:32	12:40	12:45	12:55	1:04	—	1:14
2:01	—	2:11	2:20	2:30	2:35	2:43	1:17	1:25	1:30	1:40	1:49	—	1:59
2:46	—	2:56	3:05	3:15	3:20	3:28	2:02	2:10	2:15	2:25	2:34	—	2:44
							2:47	2:55	3:00	3:10	3:19	—	3:29
3:31	—	3:41	3:50	4:00	4:05	4:13							
4:16	—	4:26	4:35	4:45	4:50	4:58	3:32	3:40	3:45	3:55	4:04	—	4:14
5:16	—	5:26	5:35	5:45	5:50	—	4:32	4:40	4:45	4:55	5:04	—	5:14
6:16	—	6:26	6:35	6:45	6:50	6:58	5:32	5:40	5:45	5:55	6:04	—	6:14
—	7:42	7:44	7:51	8:01	8:04	8:11	7:13	7:20	7:23	8:32	7:40	7:42	—
—	8:42	8:44	8:51	9:01	9:04	9:11	8:13	8:20	8:23	8:32	8:40	8:42	—
—	9:42	9:44	9:51	10:01	10:04	—	9:13	9:20	9:23	9:32	9:40	9:42	—

On the "Eastbound" schedule, the times show when each bus leaves the Crossroads mall and arrives at each of its stops. The "Westbound" schedule shows when each bus will stop on its way to the Crossroads mall.

EXAMPLE   **What time does the 12:31 Eastbound bus arrive at the Cuming Stop?**

STEP 1   Find the departing time "12:31" in the Eastbound schedule under Crossroads.   *It is in the ninth row.*

STEP 2   Go straight across to the Cuming column.   *It is the fifth column.*

STEP 3   Read the time.   **1:00**

ANSWER: The 12:31 eastbound bus will arrive at Cuming at **1:00**.

Solve each problem.

1. What time does the 6:16 Eastbound Crossroads bus reach the 61-Maple stop?

2. What time does the 8:01 Eastbound Crossroads bus reach the Burt stop?

3. What time does the Eastbound bus leave the Crossroads to reach the Jackson stop at 11:43?

4. What time does the Eastbound bus leave the Crossroads to reach the 90-Maple stop at 6:26?

5. To reach Crossroads at 9:29, what time does the Westbound bus stop at Cuming?

6. The 11:02 Jackson pickup by the Westbound bus arrives at the Crossroads at what time?

7. Brandy lives at 90th and Maple. She has an interview for a job at 23rd and Burt at 2:30 P.M. What time does she need to catch the bus at her home in order to make her interview on time?

8. Brandy's interview lasts one hour. What time will she get home if she takes the next available bus from 23rd and Burt?

9. The buses do not go to Crossroads after 6:14 P.M. What is the last pickup time at 90-Maple to get to Crossroads by 6:14 P.M.?

10. Jerry lives on Jackson Street and works at a mall called Crossroads. He must be at work at 9:00 A.M. What time will he need to catch the bus to Crossroads to be at work by 9:00 A.M.?

11. What time will he reach Crossroads?

12. How much time will the trip to work take?

13. Jerry gets off work at 6 P.M. What time will the Eastbound bus leave for Jackson Street?

14. What time will Jerry arrive at home?

15. If Jerry works late (after 6:30 P.M.), will he be able to take the Crossroads bus home?

16. The Westbound bus stops at 61-Maple at 3:55. What time does it arrive at Crossroads?

17. To reach Crossroads at 5:14, what time should you get on the Westbound bus at the 23-Burt stop?

18. On the Westbound bus, what would be the earliest and latest time that someone could arrive at the Crossroads stop?

Trucking and Delivery Service

	Atlanta, GA	Chicago, IL	Dallas, TX	Denver, CO	Detroit, MI	Kansas City, MO	Los Angeles, CA	Louisville, KY	Memphis, TN	Milwaukee, WI	Minneapolis, MN
United States Mileage Chart For Selected Cities											
Atlanta, GA		708	822	1430	732	822	2191	415	382	799	1121
Chicago, IL	708		921	1021	279	542	2048	297	537	90	410
Dallas, TX	822	921		784	1156	505	1399	828	454	1015	949
Denver, CO	1430	1021	784		1283	606	1031	1119	1043	1038	920
Detroit, MI	732	279	1156	1283		769	2288	382	719	360	685
Kansas City, MO	822	542	505	606	769		1577	513	482	564	443
Los Angeles, CA	2191	2048	1399	1031	2288	1577		2182	1807	2069	1857
Louisville, KY	415	297	828	1119	382	513	2182		378	382	705
Memphis, TN	382	537	454	1043	719	482	1807	378		622	914
Milwaukee, WI	799	90	1015	1038	360	564	2069	382	622		337
Minneapolis, MN	1121	410	949	920	685	443	1857	705	914	337	

Use the formula $r = \frac{d}{t}$, where r is velocity, d is distance, and t is time. Use the mileage chart above to answer the following questions. Round to the nearest whole number.

1. Joe drives a furniture truck around town. He spends 6 hours each day on the road and averages 45 mph. How many miles does he drive on an average day?

2. The distance between Atlanta, GA, and Minneapolis, MN, is 1,121 miles. If a truck driver averages 60 mph, how long would it take her to travel between the two cities?

3. A truck driver picked up a load in Dallas, TX, and drove 8 hours to Kansas City, MO. What was the average velocity of the truck?

4. A long-distance hauler picked up a load of oranges in Los Angeles, CA, and drove to Denver, CO. Her average speed was 65 mph. She spent 6 hours sleeping during the trip and 3 hours on breaks and meals. How many hours did the trip take?

5. Jeremy drove a round trip, short-haul load between Chicago, IL, and Milwaukee, WI. Including a half-hour break, he was away from the office for 4 hours. What was his average driving speed?

6. Mary and Joe decided to move from Detroit, MI, to Los Angeles, CA. They packed their belongings into a rental van and left Detroit at 8:00 A.M. on Saturday. They drove straight through averaging 55 mph, including breaks and other stops. How long did it take them to drive to California? What time was it in California when they arrived? (Use the Time Zones chart on page 142.)

7. How long would it take Janice to drive from Louisville, KY, to Kansas City, MO, if she averages 50 mph?

8. A trucker carried a load from Dallas, TX, to Detroit, MI. If the actual driving time was 22 hours, what was the average velocity?

9. Jack wanted to fly from Denver, CO, to Atlanta, GA. If the plane averages 350 mph, how long is the flying time?

10. If Carla averages 60 mph on a drive between Milwaukee, WI, and Detroit, MI, how long will she be on the road?

Speed of Planes

MILEAGE BETWEEN U.S. CITIES

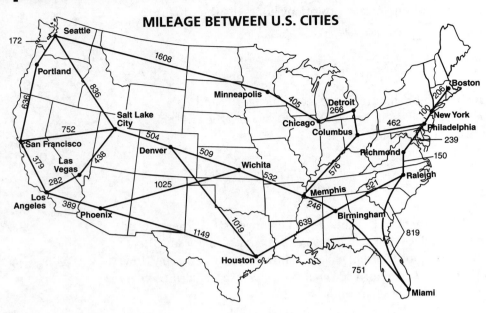

Use the formula $r = \frac{d}{t}$, where r is velocity, d is distance, and t is time. Use the mileage chart above to answer the following questions.

1. A jet picked up passengers in Denver, CO, and flew at an average velocity of 400 mph to Houston, TX. How long was the plane in the air?

2. A private plane picked up passengers in Denver, CO, and flew to Houston, TX. This plane averaged 125 mph. How long did the private plane take to reach Houston?

3. A jet flew approximately 2,000 miles from Seattle, WA, to Chicago, IL. The jet took 5 hours to make the trip. How fast was it flying?

4. The distance between New York City, NY, and Philadelphia, PA is 100 miles. How long would it take to fly between the two cities if the plane's speed is 125 mph?

5. Ann flew from Phoenix, AZ, to Wichita, KS in three hours. What was the speed of the plane?

6. Joe left Los Angeles, CA, for Seattle, WA. The plane flew at a speed of 365 miles per hour. How long did his flight take?

7. Joe took a smaller plane back home to Los Angeles from Seattle. He had a layover of one hour in Portland, OR, and his entire trip took seven hours. What was the average speed of the plane?

8. Heidi flew from Miami, FL, to Boston, MA. On the trip she stopped in Richmond, VA, where she had a two-hour layover. Her plane averaged 400 mph. How long did it take her to arrive in Boston?

9. If it takes $1\frac{1}{4}$ hours to fly from Columbus, OH, to Memphis, TN, what is the average velocity of the plane?

10. A helicopter made a special one-way flight from Chicago, IL, to Detroit, MI. If the trip took $2\frac{1}{2}$ hours, what was the speed of the helicopter? (Round your answer to the nearest whole number.)

11. A delivery plane started in Salt Lake City, UT, and then stopped in Denver, CO, and Wichita, KS, before ending the route in Memphis, TN. If the plane traveled about 200 mph, how long was the pilot in the air? (Round your answer to the nearest whole number.)

12. If it takes about 4 hours to fly from Minneapolis, MN, to Seattle, WA, what speed is the plane flying?

Posttest

Measurement Skills Review

The purpose of this review is to see how well you have mastered the skills in this measurement book. Take your time and work each problem carefully.

Rewrite each measurement so it uses the given unit.

1. 1 lb = _____ oz

2. 2.5 T = _____ lb

3. 3,500 mg = _____ g

4. 4.7 cm = _____ mm

5. 84 in. = _____ yd

6. 1 mi = _____ ft

7. 3.2 kL = _____ L

8. $6\frac{1}{2}$ qt = _____ c

9. 475 mL = _____ L

10. $3\frac{1}{4}$ hr = _____ min

11. 1 yr = _____ days

12. 287 days = _____ wk

Write the amount shown by each scale.

13.

16.

14.

17.

15.

Add, subtract, multiply, or divide.

18. 2 lb 4 oz + 3 lb 11 oz =

19. 7 lb 2 oz − 3 lb 7 oz =

20. 10 g + 400 mg =

21. 4.8 cm × 7 =

22. 1,250 km ÷ 5 =

23. 6 yd 2 ft + 4 yd 2 ft =

24. 5 qt 1 c − 2 qt 3 c =

25. $6\frac{1}{2}$ gal × 4 =

26. 275 cm ÷ 2.5 =

27. 4 days 18 hours + 2 days 20 hours =

28. 7 weeks 5 days ÷ 9 =

29. 12 hr 14 min − 7 hr 36 min =

30. 4 ft 3 in. × 10 =

For problems 31–33, use the art below.

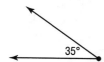

35°

31. What kind of angle is this?

_____

32. What is the measure of its complement?

_____

33. What is the measure of its supplement?

_____

For problems 34–35, use the art below.

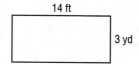

14 ft

3 yd

34. What is the perimeter?

_____

35. What is the area?

_____

36. Frank is a landscaper. He wants to make a circular flower bed with a diameter of 10 feet. He will use decorative stones as a border. To decide how many stones to buy, he needs to know the circumference. What is the circumference of the flower bed?

37. Frank wants to use fertilizer before planting the flower bed. The directions on the fertilizer bag say how much to use, based on area. What is the area of the flower bed?

38. Frank also wants to add 6 inches of top soil before planting. What is the volume of top soil he needs?

39. Veronica is a private duty nurse, caring for an elderly patient. If the patient's temperature rises 4°F or more above normal, Veronica must call the doctor. The patient's temperature is 102.4°F. Normal temperature is 98.6°F. Does Veronica need to call the doctor?

40. After having the flu for a few days, Carla found she had lost $6\frac{1}{2}$ pounds. If she now weighs 134.8 pounds, how much did she weigh before she had the flu?

41. Ryan works as a lineman for a cable television company. He cut $38\frac{1}{2}$ yards of cable from a new 200-yard spool. He needs to know if he has enough cable left to finish his jobs for the day. How much cable is left on the spool?

42. Chris and Hank are bicycle tour guides. Some of their customers are familiar with customary units and some are familiar with metric units. One week, Chris advertises a 10-mile tour, while Hank advertises a 10-kilometer tour. Whose tour is longer?

43. Tran is a home appliance repairer. He repaired eight appliances, using $22\frac{1}{2}$ centimeters of copper wire for each repair. He started with a full 20-meter spool of wire. How much wire is left on the spool?

44. Judi is a personal health and wellness coach. She keeps track of how much fat her clients eat. One of her clients ate 2 blueberry muffins each morning for four days in a row. Each muffin had $6\frac{1}{2}$ grams of fat. How many grams of fat did the client eat?

45. Mario is a sales consultant for a cement contractor. The director of a local art museum contacts Mario and describes a plan for a lawn display that will include a large cement cone with a length of 12 meters and a diameter of 8 meters. The director asks how much cement will be needed to make the cone. What volume should Mario quote?

46. A civil engineer is designing a bridge. The design includes a concrete support in the shape of a cube with edges 10 feet long. What is the volume of the concrete needed for the cube?

47. Manuel, a contractor, hired a temporary carpenter at $18.20 per hour. The carpenter worked from 7:15 A.M. to 12:30 P.M. on Monday and from 7:30 A.M. to 1:45 P.M. on Tuesday. How much does Manuel owe the carpenter for his work?

48. Carl is an independent truck driver. He keeps a record of total distance and average speed for each job. His odometer reading for one job was 28,472.6 miles at the start and 28,875.2 miles at the end. How far had he driven? If he had driven for 6 hours, what was his average speed?

49. Kai manages a temporary staffing agency. She pays her employees time-and-a-half for each hour over 40 hours that they work in a week. One employee worked $8\frac{1}{2}$ hours each day for five days at a base pay rate of $15.40 per hour. How much overtime pay does Kai pay the employee? What is the employee's total pay for the week?

50. A cargo plane averages 11.6 miles per gallon of fuel. How many gallons of fuel are needed for a trip of 1,485 miles? (Round your answer to the nearest gallon.)

SKILLS REVIEW CHART

If you had fewer than 40 correct, review the chapters where you missed problems. Rework any problems you missed. Here is a list of the problems and where each skill is covered.

Problem Numbers	Chapter
4–6, 14, 21–23, 26, 30–36, 41–43	Length and Angles
1–3, 13, 16, 18–20, 39, 40, 44	Weight and Temperature
7–9, 15, 24, 25, 37, 38, 45, 46	Capacity and Volume
10–12, 17, 27–29, 47–50	Time and Velocity

ANSWER KEY

Skills Inventory Answers Pages 1–4

1. 8,789
2. 598,977
3. 111
4. 8,223
5. 6,879
6. 245
7. 3,031
8. 8,078
9. 2,093
10. 143,165
11. 159
12. 368
13. 24,408
14. 945
15. 10,926
16. 51,285
17. 11,251,275
18. 123,800
19. 347,000
20. 35
21. 420
22. 578 r 6 or 578.75
23. 176 r 8 or 176.29
24. 119 r 20 or 119.5
25. 4.67
26. $\frac{3}{4}$
27. $\frac{1}{2}$
28. $\frac{5}{8}$
29. $5\frac{5}{6}$
30. $6\frac{1}{4}$
31. $\frac{3}{5}$

32. $\frac{1}{6}$
33. $2\frac{1}{3}$
34. $2\frac{9}{10}$
35. $\frac{7}{20}$
36. 15
37. $\frac{3}{2} = 1\frac{1}{2}$
38. $1\frac{3}{5}$
39. 5
40. $\frac{5}{8}$
41. $5\frac{1}{5}$
42. 7.4
43. 12.93
44. 82.522
45. 3.21
46. 157.85
47. 348.86
48. 834.3
49. 0.376
50. 4,256
51. 1,290.51
52. 21
53. 350
54. 17.2
55. 0.8475
56. $155.50
57. $21,320
58. 190 lb
59. 22 yards
60. $2\frac{3}{4}$ gallons
61. 20 cloths
62. $273.09
63. 5.2 lb

Chapter 1 Pages 8–9

1. yd
2. T
3. in.
4. pt
5. mi
6. fl oz
7. lb
8. gal
9. qt
10. 8
11. 12
12. 2,000
13. 16
14. 4
15. 48
16. 2
17. 1,760
18. 12
19. Sample: car trip, distance between cities, distance across an ocean
20. Sample: milk, gas, paint, water
21. Sample: candy, fruit, rice
22. Sample: meat, body weight, packages
23. mL
24. kg
25. m
26. km
27. L
28. mg
29. g
30. mm
31. kL
32. 1,000
33. 1,000
34. 1,000
35. 0.001
36. 1,000
37. 1,000
38. Sample: races at a track meet, wallpaper, fabric (outside U.S.)
39. Sample: distance between cities (Europe), distance in marathons, reading on odometer
40. Sample: medications, fat, cholesterol, protein
41. Sample: bottled water, soft drinks

Page 11

1. 44°F
2. 85°F
3. 32°F
4. 12°C
5.

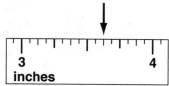

6.

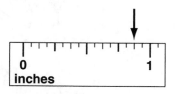

7.

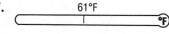

8.

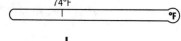

9.

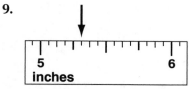

10.

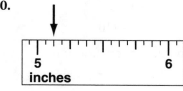

11.

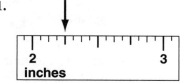

12.

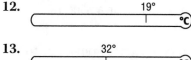

13.

14.

Pages 12–13

1. 2,301 kWh
2. 5,732 kWh
3. 9,905 kWh

4.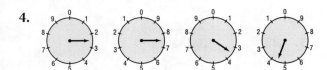

5.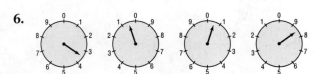

6.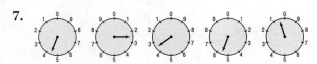

7.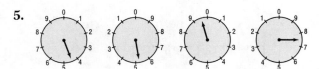

8.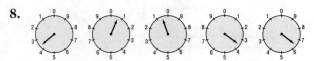

Pages 14–15

1–11. Answers will vary.

12. Area = 240 sq ft, perimeter = 62 ft,
$A = \ell \times w, P = 2\ell + 2w$

13. Area = 146, Perimeter = 110

14. No; Yes (if the outline is very uneven)

Chapter 2 Page 17

1. Sample for mm: width of a splinter, width of a pin, screw in eye glasses

2. Sample for cm: width of clear tape, thickness of a notepad, width of a pencil

3. Sample for m: width of a door, a window, length of a towel

4. Sample for km: race, distance from home to school, 20 laps in swimming pool

5. 1.3 cm

6. 3.9 cm

7. 4.3 cm

8. 93.4 cm

9. 96 cm

10. 99.8 cm

11.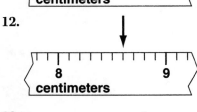

12.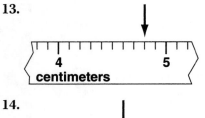

13.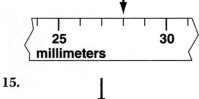

14.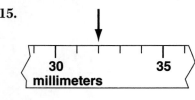

15.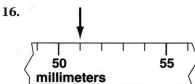

16.

Pages 18–19

1.	2	**12.**	0.188
2.	10,000	**13.**	300
3.	7.5	**14.**	5,290
4.	30	**15.**	2,700
5.	0.6	**16.**	21,000
6.	1,000	**17.**	700,000
7.	2.2	**18.**	12,000
8.	6.25	**19.**	2,250
9.	147.5	**20.**	32
10.	8.7	**21.**	40,500
11.	4.9	**22.**	1,600

23. 127.5   **27.** 235 mm

24. 5,200   **28.** 21,680 m

25. 820,000   **29.** 5.8 m

26. 360   **30.** 3.5 km

Page 20

1. 202.5 m   **5.** 3.3003 km

2. 12.81 km   **6.** 2,387 cm

3. 11.6 mm   **7.** 0.14 km

4. 43 cm   **8.** 3 m

Pages 21–23

1. Sample: rainfall, snowfall, nails, bolts

2. Sample: swimming pool depth, construction materials such as plywood and electrical wire, distances in parking rules, height of a person

3. Sample: carpet, distances in a football game, cloth material, length of yarn or string

4. Sample: distance between cities, speed, odometer readings, distance a person jogs

5. jump rope

6. hammer

7. extension cord

8. screwdriver

9. 38 in.

10. 100 yd

11. 42 in.

12. 86 in.

13. 18 ft

14. 60 ft

15. 325 mi

16. 50 ft

17. 25 yd

18. 410 mi

19. a pair of pliers

20. a child

21. a roll of adhesive tape

22. a tricycle

23. a dish towel

24. $\frac{1}{2}$ inch

25. $1\frac{1}{4}$ inch

26. $2\frac{3}{4}$ inch

27. $6\frac{3}{4}$ inch

28. $8\frac{1}{8}$ inch

29. $9\frac{3}{8}$ inch

30.

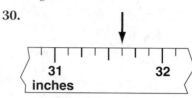

31.

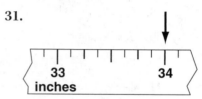

32.

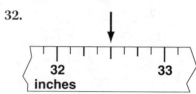

33.

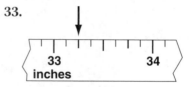

34.

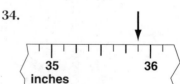

35.

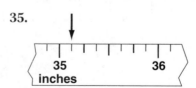

Pages 24–25

1. 108   **5.** 21,120

2. 72   **6.** 35,200

3. 30   **7.** 84

4. 360   **8.** 52,800

9. 5,280

10. 66

11. $19\frac{1}{2}$

12. 18,040

13. 2

14. 2

15. 11

16. $4\frac{2}{3}$

17. $80\frac{2}{3}$

18. $\frac{1}{2}$

19. 11,088

20. 102

21. 259.2

22. 3

23. 6

24. 2

25. $2\frac{1}{6}$

26. $2\frac{1}{2}$

27. $1\frac{1}{3}$

28. 6,600

29. 37

30. 78

31. 2

32. 20

33. 5

34. $2\frac{1}{3}$

35. 1.5

36. $2\frac{1}{6}$

37. 7 ft

38. 68,640 ft

39. Yes.

40. 80 yd

41. 198

42. 42 in.

43. Yes.

Pages 26–28

1. 10 yd 1 ft

2. 10 ft 3 in.

3. 8 yd 3 in.

4. 9 ft 6 in.

5. 5 yd 18 in.

6. 7 yd 8 in.

7. 4 yd 2 ft

8. 1 ft 8 in.

9. 1 ft 5 in.

10. 5 yd 1 ft 7 in.

11. 2 yd 2 ft

12. 9 yd 11 in.

13. 4 yd 1 in.

14. 7 yd 1 ft

15. 3 ft 10 in.

16. 6 yd 2 ft 4 in.

17. 1 ft 7 in.

18. 7 yd 2 ft 10 in.

19. 3 ft 8 in.

20. 24 ft 6 in.

Page 30

1. 31 ft 8 in.

2. 3 yd 1 ft

3. 3 ft 5 in.

4. 6 yd

5. 1 yd 1 ft

6. 4 in.

7. 2 yd 1 ft

8. 1 ft 11 in.

9. 1 ft 3 in.

10. 32 ft

11. 17 ft 6 in.

12. 2 ft 4 in.

13. 3 yd

14. 2 ft 1 in.

15. 53 yd 1 ft

16. 21 yd 1 ft 2 in.

17. 30 ft

18. 18 ft 8 in.

19. 1 ft 7 in.

20. 1 ft 7 in.

21. 0

22. 2 ft 7 in.

23. 12 yd 2 ft 8 in.

24. 16 ft 8 in.

25. 2 ft 6 in.

Page 31

1. 1 in.

2. 1 km

3. 1 yd

4. 1 m

5. 1 ft

6. 1 km

7. 1 mi

8. 1 km

9. 1 m   **14.** 10 cm

10. 25 m   **15.** 15 cm

11. 3 in.   **16.** 28 in.

12. 60 in.   **17.** 3 yd

13. 1 yd

Pages 32–33

1. 3 mm   **11.** >

2. 60 in.   **12.** =

3. 65 m   **13.** =

4. 2 yd   **14.** <

5. 2,400 m   **15.** =

6. 48 in.   **16.** >

7. 570 cm   **17.** <

8. 10,000 ft   **18.** >

9. 3 m   **19.** <

10. <

20. 2 mi, 2 yd, 2 ft, 2 in.

21. 3 m, 35 cm, 67 mm, 6 cm

22. 3 mi, 26 yd, 18 ft, 4 yd

23. 4 m, 350 cm, 6 cm, 20 mm

24. Yes.

25. Yes.

26. No.

27. Yes.

Pages 34–35

1. ∠B, ∠ABC, or ∠CBA

2. ∠PXS or ∠SXP

3. ∠PXT or ∠TXP

4. ∠TXR or ∠RXT

5. ∠RXS or ∠SXR

6. ∠EHD or ∠DHE

7. ∠EHF or ∠FHE

8. ∠GHF or ∠FHG

9. 130°

10. 45°

11. 90°

12.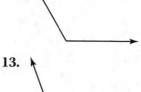

13.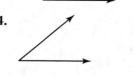

14.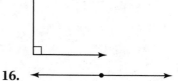

15.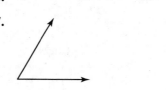

16.

17.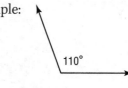

Page 36

1. acute, 40°

2. acute, 80°

3. acute, 87°

4. obtuse, 110°

5. obtuse, 165°

6. right, 90°

7. sample:

80°

8. sample:

110°

9.

10.

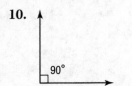

Page 37

1. 30°	**6.** 30°
2. 150°	**7.** 150°
3. 30°	**8.** 180°
4. 30°	**9.** 180°
5. 150°	

Pages 38–39

1. 15°

2. 59°

3. 22°

4. 43°

5. 51°

6. 4°

7. 55°

8. 5°

9. 100°

10. 60°

11. 140°

12. 155°

13. 35°

14. 112°

15. 90°, right

16. 140°, obtuse

17. 60°, acute

18. 40°, acute

19. 95°, obtuse

20. 160°, obtuse

Page 41

1. $P = 28$ cm, $A = 48$ sq cm

2. $P = 34$ ft, $A = 42$ sq ft

3. $P = 36$ cm, $A = 45$ sq cm

4. $P = 28$ cm, $A = 49$ sq cm

5. $P = 7$ ft, $A = 3$ sq ft

6. $P = 51.4$ cm, $A = 164.7$ sq cm

7. $P = 8\frac{2}{3}$ yd, $A = 4$ sq yd

8. $P = 42$ cm, $A = 20$ sq cm

9. $P = 3.09$ m, $A = 0.414$ sq m

10. 96 feet

11. 560 tiles

12. 3 cans

13. $C \approx 94.2$ cm, $A \approx 706.5$ sq cm

Pages 42–44

1. 36

2. 5,280

3. 10

4. 180

5. 12

6. 100

7. 90

8. 1,000

9. 3

10. 60

11.

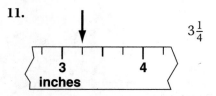

$3\frac{1}{4}$

12.

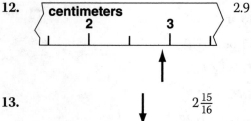

2.9

13.

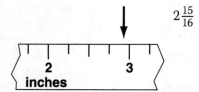

$2\frac{15}{16}$

14. 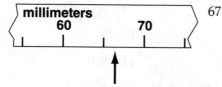 67

15. 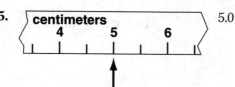 5.0

16. 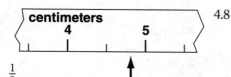 4.8

17. $\frac{1}{2}$

18. 600

19. 65

20. $7\frac{1}{2}$

21. 42

22. $25\frac{1}{2}$

23. 3

24. $2\frac{1}{2}$

25. 2.5

26. 325

27. 2 yd, 1 yd 6 in., 38 in., 3 ft

28. 80 m, 18 m, 1,600 cm, 18 cm

29. 10 ft 3 in.

30. 2 yd 7 in.

31. 0.5 mi

32. 36 mi

33. acute, 65°

34. obtuse, 155°

35. right, 90°

36. straight, 180°

37. 28°

38. 73°

39. 10°

40. 75°

41. 89°

42. 45°

43. 65°

44. 140°

45. 60°

46. 170°

47. 5°

48. 165°

49. 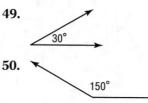 30°

50. 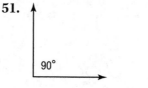 150°

51. 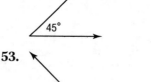 90°

52. 45°

53. 135°

54. 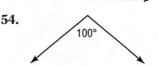 100°

55. $P = 34$ cm, $A = 70$ sq cm

56. $C \approx 31.4$ ft, $A \approx 78.5$ sq ft

57. 310 ft

58. 28 yd

Chapter 3 Pages 46–48

1. Sample: dose of liquid medicine, cream to put in coffee, a dozen peanuts

2. Sample: ground beef for dinner, a can of mixed vegetables, a medium size soft drink

3. Sample: a load of firewood, mulch for a garden, a large horse

4. a melon - heaviest, a raspberry - lightest

5. a garbage truck - heaviest, a motorcycle - lightest

6. a gallon of milk - heaviest,
a glass of juice - lightest

7. a textbook - heaviest, a paper clip - lightest

8. a 46-inch television - heaviest,
an MP3 player - lightest

9. 1 lb 8 oz

10. 13 oz

11. 1 lb 15 oz

12.

13.

14.

15. $5\frac{1}{2}$ oz

16. $3\frac{1}{2}$ oz

17. $6\frac{1}{2}$ oz

18.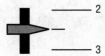

19.

20.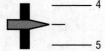

Page 49

1. mg; Sample: a hair, a pin, a thread

2. g; Sample: a pain tablet, protein, packaged food

3. kg; Sample: body weight, package, cargo

4. 400 g

5. 3 kg

6. 200 mg

7. 15 kg

8. 40 kg

9. 30 g

10. 220 kg

11. 800 g

12. 100 mg

13. 400 kg

Page 51

1. 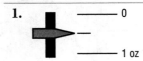 $\frac{1}{2}$ oz

2. 250 lb

3. 1,000 lb

4. $3\frac{1}{2}$ lb

5. 5 lb

6. 450 lb

7. $6\frac{1}{2}$ oz

8. 5 oz

9.  7 lb 0 oz

10. 3,000 lb

11. $2\frac{1}{2}$ lb

12.  20 lb

Pages 52–53

1. 96	**19.** 2
2. 10,000	**20.** 10
3. 32	**21.** 1
4. 4,000	**22.** 4
5. 6,000	**23.** 5
6. 160	**24.** 7
7. 40,000	**25.** 12
8. 288	**26.** 20
9. 64	**27.** 20
10. $1\frac{1}{4}$	**28.** 20
11. $5\frac{1}{4}$	**29.** 18,400
12. $1\frac{3}{8}$	**30.** 4,200
13. $3\frac{2}{5}$	**31.** 9,000
14. $1\frac{4}{5}$	**32.** 36.8
15. $\frac{3}{4}$	**33.** 6,500
16. $4\frac{1}{4}$	**34.** 104
17. $\frac{1}{2}$	**35.** 44
18. $\frac{1}{2}$	**36.** 88

37. 5,200 lb	**42.** 44,000 lb
38. 200 oz	**43.** 18 T
39. 1.4 T	**44.** 28 oz
40. $1\frac{3}{8}$ lb	**45.** $7\frac{1}{2}$ lb
41. 24 four-oz patties	**46.** 5 lb 10 oz

Page 54

1. g	**13.** kg
2. kg	**14.** g
3. g	**15.** kg
4. kg	**16.** kg
5. g	**17.** g
6. g	**18.** g
7. kg	**19.** mg
8. g	**20.** kg
9. kg	**21.** g
10. g	**22.** kg
11. kg	**23.** kg
12. kg	

Pages 55–57

1. 6,000	**15.** 20
2. 27,000	**16.** 6
3. 18,000	**17.** 7
4. 17,000	**18.** 8
5. 245,000	**19.** 2
6. 3,000	**20.** 26
7. 3,600	**21.** 2.1
8. 8,600	**22.** 4.725
9. 6,300	**23.** 2.75
10. 4,250	**24.** 4.2
11. 12,600	**25.** 0.325
12. 2,450	**26.** 3.6
13. 6,400	**27.** 3,500 mg
14. 7,250	**28.** 9,100 g

29. 90,000 mg

30. 1,500 mg

31. eight 500-gram packages

32. 1,750 g

33. 4.7 g

34. 0.1 kg

35. 56 g

36. 1.75 kg

37. 3.5 kg

38. 32.5 g

Page 59

1. 11 lb 4 oz

2. 1 lb 6 oz

3. 25 lb 5 oz

4. 4 lb 8 oz

5. 4 lb 5 oz

6. 6 lb 10 oz

7. 11 lb 5 oz

8. 10 oz

9. 5 lb 14 oz

10. 16 lb 5 oz

11. 4 lb 9 oz

12. 4 lb 5 oz

13. 2,036 g or 2.036 kg

14. 14,600 mg or 14.6 g

15. 11.25 kg or 11,250 g

16. 12,012 mg or 12.012 g

17. 1.001001 kg or 1,001.001 g or 1,001,001 mg

18. 3 kg or 3,000 g or 3,000,000 mg

19. 12 lb 8 oz

20. Yes.

Page 61

1. 6 lb 15 oz

2. 1 lb 7 oz

3. 9 lb 8 oz

4. 3 lb 1 oz

5. 2 lb 7 oz

6. 30 lb 8 oz

7. 1 lb 8 oz

8. 1 lb 12 oz

9. 73 lb 14 oz

10. 3 lb 11 oz

11. 97 lb 12 oz

12. 5 lb 7 oz

13. 30 lb 3 oz

14. 1 lb 2 oz

15. 8 lb 2 oz

16. 15 lb 5 oz

17. 4 oz

18. 5 lb 10 oz

Pages 62–63

1. 3 lb

2. 3.5 kg

3. $2\frac{1}{2}$ lb

4. 4,800 lb

5. $7\frac{1}{2}$ lb

6. 2.5 g

7. 18 oz

8. 360 g

9. <

10. >

11. <

12. =

13. <

14. >

15. >

16. =

17. 7 oz, 14 oz, 1.5 lb, 2 lb

18. 18 lb, 270 lb, 1.6 T, 3 T

19. 16 mg, 6 g, 482 g, 6 kg

20. 240 mg, 45 g, 800 g, 1 kg

21. the 3-lb package

22. Yes.

23. 22 oz is greater.

24. 400 mg is smaller.

Pages 64–65

1. 1 lb

2. 1 ton

3. 1 kg

4. 1 oz

5. 1 lb

6. 1 ton

7. 1 kg

8. 1 g

9. 1 ton

10. 1 oz

11. Sample: an ice cube, a hair curler, a hard candy

12. Sample: a car, a load of decorative rock, a cement wall

13. Sample: a cake, a gallon of milk, a textbook

14. Sample: a bag of flour, a basket of apples, a water jug

15. food labels

16. 6 g

17. 3 kg

18. 1 lb

19. 24 kg

20. 1 ton

21. 400 mg

22. $\frac{1}{2}$

23. less

24. more

25. less

26. $2\frac{3}{10}$

27. 909

Pages 66–67

1. 37°C; 98.6°F   **14.** 70°F
2. 100°C; 212°F   **15.** 68°F
3. 100°C; 180°F   **16.** 80°C
4. 86°F   **17.** 28°F
5. 176°F   **18.** 8°C
6. 40°C   **19.** 98.6°F
7. 90°C   **20.** 12°F
8. 122°F   **21.** 6°F
9. 28°C   **22.** 0°F
10. 90°C   **23.** 29°F
11. 24°C   **24.** 150°F
12. 48°C   **25.** 104°F
13. 30°F   **26.** 20°C

27. 70° °F
28. 180° °F
29. 65° °C
30. 42° °C

Pages 68–69

1. 60°F   **11.** 3°F
2. 48°C   **12.** −15°F
3. 19°F   **13.** −3°F
4. 19°C   **14.** −4°C
5. 86°F   **15.** −8°F
6. 17°C   **16.** −16°C
7. −3°C   **17.** 32°C
8. −7°C   **18.** 0°F
9. −6°F   **19.** −7°C
10. 15°C   **20.** 30°F

21. 36° °F
22. −12° °F

23. 48° °F
24. 0° °F
25. 45° °F
26. −5° °F
27. −17° °F

Pages 70–71

1. 86°F   **13.** −20°C
2. 104°F   **14.** 15°C
3. 167°F   **15.** 100°C
4. 50°F   **16.** −40°C
5. 113°F   **17.** 212°F
6. 194°F   **18.** 32°F
7. 40°C   **19.** −5°C
8. 70°C   **20.** 25°C
9. 20°C   **21.** 20°C
10. 0°C   **22.** 40°C
11. 22.2°C   **23.** 5°C
12. 30°C   **24.** 86°F

Pages 72–75

1. 16   **12.** 32°C
2. 2.2   **13.** 600 g
3. 1,000   **14.** 0°C
4. 2,000   **15.** $2\frac{1}{2}$ T
5. 28   **16.** 15 oz
6. $4\frac{1}{8}$ **17.** 80°F
7. $2\frac{1}{4}$ **18.** 16 oz
8. 3.5   **19.** 26°F
9. 1,120   **20.** 1 kg
10. 4,300   **21.** 2.2 T
11. 2 kg

22.

23.

24.

25. 26 g, 26 oz, 26 lb, 26 kg

26. 2 oz, 2 lb, 2 T

27. 1 mg, 1 g, 1 kg

28. $2\frac{1}{4}$ lb

29. 5 lb 11 oz

30. 1 lb 14 oz

31. 5 lb 4 oz

32. 98.6°F

33. 15°C

34. 14 lb 8 oz

35. 4 lb 10 oz

36. 95°F

37. 32°F or 0°C

38. 212°F or 100°C

39. 20°C

40.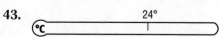

41.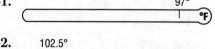

42.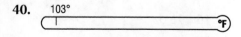

43.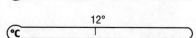

44.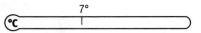

45.

46.

47. Possible answer

48. Possible answer

49.

50.

51.

52.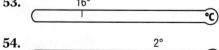

53.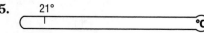

54.

55.

56.

57. Possible range: 20° to 25°C

58. Possible range: below 0°C

59. inch, foot, yard, mile

60. ounce, pound, ton

61. meter, kilometer, millimeter

62. gram, milligram, kilogram

63. 2,447

64. 29,039

65. 37°F

66. 87°F

Chapter 4 Pages 76–77

1. Sample: milk, ice cream, fruit drinks, paint

2. Sample: coffee, juice, frozen yogurt, sugar, flour

3. Sample: honey, butter, sugar

4. qt

5. c

6. qt

7. pt

8. tsp

9. c

10. gal

11. tsp

12. oil for brownies—smallest,
    oil for a car—largest

13. a picnic jug—smallest,
    swimming pool—largest

14. a cup of water—smallest,
    a gallon of milk—largest

15. vanilla—smallest,
    a bowl of punch—largest

16. salt—smallest,
    flour—largest

17. 1 c 8 tbsp

18. 2 c 0 tbsp

19. 4 tbsp

Page 79

Possible answers are given.

1.  2 c

2.  $\frac{3}{4}$ c

3. 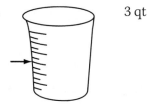 3 qt

4.  1 qt

5.  $1\frac{1}{2}$ tsp

6.  $1\frac{1}{2}$ qt

7.  2 qt

8.  $1\frac{1}{4}$ tsp

9.  1c

10. 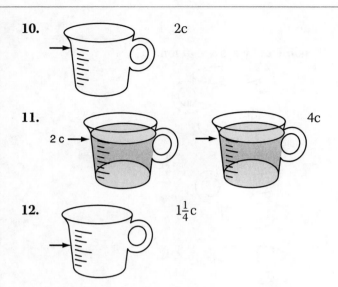 2c

11. 2 c → 4c

12. $1\frac{1}{4}$ c

Pages 80–83

1. 18	**25.** 40
2. 80	**26.** 3
3. 12	**27.** $\frac{1}{2}$
4. 32	**28.** $5\frac{1}{2}$
5. 16	**29.** $6\frac{2}{3}$
6. 16	**30.** $1\frac{1}{3}$
7. 40	**31.** $1\frac{7}{8}$
8. 24	**32.** 2
9. 50	**33.** $1\frac{3}{4}$
10. 24	**34.** $1\frac{1}{2}$
11. 29	**35.** $1\frac{9}{16}$
12. $17\frac{1}{3}$	**36.** 12 doses
13. $6\frac{3}{4}$	**37.** 30 gal
14. 15	**38.** 32 c
15. 76	**39.** 6 c
16. 26.4	**40.** 9 tsp
17. 15	**41.** 8 tbsp
18. $\frac{1}{2}$	**42.** 40
19. 2	**43.** 40
20. 5	**44.** 4 c
21. $1\frac{1}{2}$	**45.** 2 pt
22. 9	**46.** 10 qt
23. 1	**47.** 2 tbsp
24. 2	

Pages 84–85

1. Sample: medicine, blood sample, chemicals

2. Sample: soda, gasoline (Europe), milk

3. Sample: water in pool, gasoline in a tanker, water output (Europe)

4. mL	**18.** 2.5 L
5. kL	**19.** 4 L
6. L	**20.** 2 L
7. kL	**21.** 6 kL
8. L	**22.** 16 L
9. kL	**23.** 206 kL
10. mL	**24.** 16 L
11. L	**25.** 240 mL
12. mL	**26.** 25 mL
13. 16 mL	**27.** 14 kL
14. 1 L	**28.** 1 mL
15. 28 L	**29.** 1 kL
16. 2 L	**30.** milliliter, liter, kiloliter
17. 3 mL	

Pages 86–88

1. 4,000	**11.** 3,200
2. 0.003	**12.** 50
3. 0.6	**13.** 10
4. 0.01	**14.** 0.0006
5. 6,000	**15.** 0.15
6. 5,000	**16.** 0.00078
7. 1,500	**17.** 100
8. 2.4	**18.** 0.0008
9. 5,500	**19.** 6.2
10. 0.0032	**20.** 0.009225

Pages 89–90

1. 9 gal 1 qt	**3.** 17 gal 5 c
2. 3 c 13 tbsp	**4.** 5 gal 2 qt

5. 9 qt 1 c

6. 3 pt 1 c

7. 2 pt 1 c

8. 10 qt 1 pt

9. 1 qt 2 c

10. 9 pt

11. 2 qt 3 c

12. 3 gal 1 qt 1 pt

13. 2.035 kL

14. 460 mL

15. 3,507.2 L

16. 7,250 mL

17. 2,800 mL

18. 0.85 kL

19. 5 gal 1 qt

20. 1,750 mL

21. 9 L

22. 5 c

Page 92

1. 15 pt

2. 2 gal 1 qt

3. 1 L 50 mL

4. 1 kL 750 L

5. 22 gal 2 qt

6. 35 qt

7. 1 gal 2$\frac{1}{2}$ qt

8. 51.2 L

9. 0.9 kL

10. 4 L 50 mL

11. 1 qt 3 c

12. 14 gal 2 pt

13. 49 qt

14. 18 c 12 tbsp

15. 240 L 20 mL

16. 1,770 mL, 1.77 L

17. 3 qt 2 c

18. 960 mL

19. 20 loads

Page 93

1. 1 qt

2. 1 kL

3. 1 L

4. 1 fl oz

5. 1 L

6. 1 gal

7. 1 tsp

8. 1 gal

9. 1 kL

10. 1 pt

11. 1 kL

12. 1 fl oz

13. 400 mL

14. 1,000 kL

15. 6 qt

16. 1 c

17. 15 gal

18. 1 gal

19. 6 fl oz

Pages 94–97

1. 600 cu ft

2. 48.75 cu in

3. 1,555.2 cu m

4. 336 cu in.

5. Sample: amount a box holds, dirt to be moved, mulch to be put out

6. $10 \times 3 \times 5 = 150$ cu ft

7. $6 \times 0.5 \times 2 = 6$ cu m

8. $3 \times \frac{1}{3} \times 2 = 2$ cu yd

9. $2 \times 6 \times 18 = 216$ cu in.

10. 8 cu yd

11. $0.35 \times 2 \times 0.05 = 0.035$ cu m

12. Sample: sugar cube, toy blocks, storage containers

13. $2 \times 2 \times 2 = 8$ cu yd

14. 216 cu in.

15. 15.625 cu m

16. 10,648 cu mm

17. 3,375 cu in.

18. 144 cu in.

19. $\frac{1}{8}$ cu in.

20. 13.824 cu m

21. 4 ft

22. 10 m

23. 8 in.

24. 20 in.

25. 10 in.

26. 12 in. or 1 ft

27. 512 cu in.

28. 3 ft

Pages 98–101

1. $3.14 \times 2^2 \times 12 = 150.72$ cu ft

2. $3.14 \times 6^2 \times 20 = 2,260.8$ cu cm

3. 5,024 cu m

4. 502.4 cu in.

5. 147.1875 cu cm

6. 753.6 cu ft

7. 602,880 cu mm

8. 18.84 cu yd

9. 1,055.04 cu in.

10. 1,130.4 cu m

11. 18.84 cu yd

12. 4,396 cu m

13. 5,024 cu ft

14. 4,710,000 cu cm

15. $\frac{1}{3} \times 3.14 \times 5^2 \times 10 = 261.67$ cu yd

16. $\frac{1}{3} \times 3.14 \times 10^2 \times 12 = 1,256$ cu in.

17. $\frac{1}{3} \times 3.14 \times 8^2 \times 12 = 803.84$ cu cm

18. $\frac{1}{3} \times 3.14 \times 9^2 \times 9 = 763.02$ cu in.

19. $\frac{1}{3} \times 3.14 \times 3^2 \times 10 = 94.2$ cu ft

20. $\frac{1}{3} \times 3.14 \times (\frac{1}{2})^2 \times 18 = 4.71$ cu in.

21. $3.14 \times 9^2 \times 30 = 7,630.2$ cu in.
$7,630.2 \times 7.5 = 57,226.5$ gal

22. $3.14 \times 2.5^2 \times 6 = 117.75$ cu ft

23. $(785 \div 10) + 3.14 = 25$, so radius = 5 in.

24. 30 cm

25. 100.48 cu in.

Pages 102–105

1. 4

2. 1

3. 1,000

4. 3

5. $\frac{1}{2}$

6. 8

7. 0.001

8. 32

9. 4

10.

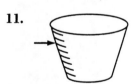

11.

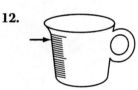

12.

13. 1 qt

14. 1 mL

15. 1 c

16. 1 L

17. 1 tsp

18. 1 pt

19. 2 fl oz

20. 8 fl oz

21. 1 c

22. 1 qt

23. 1 L

24. 3

25. $10\frac{1}{2}$

26. 16

27. 12

28. 0.75

29. 2,400

30. $4\frac{1}{2}$

31. 112

32. 1,600

33. fl oz

34. L

35. c

36. gal

37. mL

38. qt

39. 2,400

40. 6

41. 10

42. $1\frac{1}{2}$

43. 48

44. 9

45. 3

46. $3\frac{3}{4}$

47. 3 tsp, 3 c, 3 pt

48. 750 mL, 1500 mL, 7 L

49. $6\frac{1}{2}$ qt, 28 c, 14 qt

50. 2 qt

51. 1 kL

52. 4 tbsp

53. 250 mL

54. 5 qt

55. 15 gal

56. 2 qt 3 c

57. 1 gal $2\frac{1}{2}$ qt

58. 11.05 L

59. 17 pt 1 c

60. 216 cu in.

61. 282.6 cu m

62. 160 cu ft

63. $1\frac{3}{4}$ qt

64. 6 qt

65. 120 cu in.

66. 3 c

67. 12 doses

68. 20 in.

69. 56.52 cu in.

70. 0.75 L or 750 mL

71. metric - gram, milligram, kilogram
customary - ounce, pound, ton

72. metric - milliliter, liter, kiloliter
customary - teaspoon, tablespoon, cup, pint, quart, gallon

73. metric - millimeter, meter, kilometer
customary - inch, yard, foot, mile

74. $5\frac{1}{8}$ in.

75. $3\frac{5}{8}$ in.

76. $\frac{7}{8}$ in.

77. 425 mm

78. 408 mm

79. 379 mm

80. $6\frac{1}{2}$ in.

81. 35 mm

82. 27°

83. 63°

84. 65°

85. 135°

Chapter 5 Pages 106–108

1. Sample: running 100 yd dash, warm food in microwave, checking pulse

2. Sample: summer vacation, temporary job, grow garden vegetables

3. Sample: your work day, travel to another state, fix engine on car

4. years

5. hours or days

6. minutes

7. minutes

8. years

9. hours

10. minutes

11. weeks

12. 2 hours

13. 8 hours

14. 20 minutes

15. 4 weeks

16. 15 months

17. 20 minutes

18. 12 weeks

19. 40 minutes

20. 2 minutes 35 seconds

21. 0 minutes 30 seconds

22. 1 minute 55 seconds

23. 1 minute 0 seconds

24. 1 minute 35 seconds

25. 1 minute 59 seconds

26. 0 minutes 42 seconds

27. 2 minutes, 8 seconds

28. 2 minutes 2 seconds

29.

30.

31.

32.

33.

34.

Page 109

1. Sample: 20 minutes
2. Sample: 5 minutes
3. Sample: 15 minutes
4. Sample: $1\frac{1}{2}$ hours
5. Sample: 10 minutes
6. Sample: $1\frac{1}{2}$ hours
7. Sample: 3 minutes
8. Sample: $4\frac{1}{2}$ hours
9. Sample: 2 hours

10. Sample: 30 minutes
11. weeks
12. min
13. hr
14. sec
15. min
16. days
17. hr
18. hr
19. yr
20. weeks

Pages 110–113

1.	432,000	25.	5
2.	432,000	26.	4
3.	1,728	27.	24
4.	2,880	28.	$1\frac{1}{2}$
5.	240	29.	$91\frac{1}{4}$
6.	600	30.	$1\frac{1}{4}$
7.	10,080	31.	$1\frac{1}{2}$
8.	180	32.	$1\frac{3}{4}$
9.	2,700	33.	$1\frac{1}{3}$
10.	150	34.	2.5
11.	150	35.	2 days 2 hr
12.	738	36.	$1\frac{2}{3}$ hr
13.	252	37.	2,400 min
14.	60	38.	3 yr
15.	228	39.	40 hr
16.	165	40.	80 hr
17.	2,160	41.	$\frac{1}{2}$ min
18.	336	42.	8 hr
19.	60	43.	180 min
20.	2	44.	30 min
21.	2	45.	1 hr 30 min
22.	7	46.	3 min
23.	10	47.	Brand B is 20 seconds faster.
24.	4		

Pages 114–115

1. 11 yr 5 mo
2. 19 weeks 3 days
3. 23 min 21 sec

4. 3 hr 25 min

5. 4 days 18 hr

6. 5 years 8 weeks

7. 10 hr 15 min

8. 14 hr 7 min

9. 3 days 11 hr

10. 9 hr 25 min

11. 28 days 18 hr

12. 49 min 33 sec

13. 11 min 42 sec

14. 3 weeks

15. 28 days 11 hr

16. 11 hr 15 min

17. 16 yr 9 mo

18. 12 min 10 sec

19. 34 hr 30 min

20. 3 hr 21 min

Pages 116–117

1. 1 hr 14 min

2. 2 weeks 2 days

3. 2 mo 5 days

4. 9 hr 16 min

5. 9 yr 4 mo

6. 76 min 40 sec

7. 1 yr 10 mo

8. 20 hr

9. 49 yr

10. 18 mo 9 days

11. 32 days 22 hr

12. 4 days 6 hr

13. 33 min 10 sec

14. 28 hr 48 min

15. 2 hr 13 min 20 sec

16. 6 min 3 sec

17. 3 yr 5 mo

18. 26 hr 15 min

19. $17\frac{1}{2}$ hr

Pages 118–119

1. 80 hours

2. 360 minutes

3. 3 months

4. 2 years

5. 30,000 sec

6. $3\frac{1}{2}$ days

7. 274 days

8. 50 days

9. 160 weeks

10. 248 minutes

11. 2 months

12. $3\frac{1}{2}$ years

13. =   19. <

14. <   20. >

15. >   21. =

16. >   22. <

17. >   23. <

18. <   24. =

25. 45 seconds, 3 minutes, 200 seconds

26. 420 minutes, $2\frac{1}{2}$ days, 65 hours

27. 36 weeks, 10 months, 1 year

28. 85 weeks, $2\frac{1}{2}$ years, 40 months

29. Jose

30. less

31. less

32. same

33. Masao

Pages 120–121

1. 2.5 mph   4. 9,000 mi

2. 300 mph   5. 65 mi

3. 120 mph   6. 165 mi

7. 6 hr	**14.** 520 mi	**41.** 8 hr
8. 300 mph	**15.** 12 hr	**42.** No.
9. 270 mi	**16.** 400 mph	**43.** 8 hr 12 min
10. 2 hr	**17.** $13\frac{3}{4}$ mph	**44.** 8 hr
11. $2\frac{1}{4}$ hr	**18.** 60 mph	
12. 360 mph	**19.** 3.5 hr	
13. 5.5 hr		

41. 8 hr

42. No.

43. 8 hr 12 min

44. 8 hr

45. Yes.

46. 12 hr

47. 6:15 A.M.

Pages 122–125

1. 60	**14.** $2\frac{1}{3}$
2. 12	**15.** 21
3. 24	**16.** $4\frac{1}{2}$
4. 4	**17.** 390 min
5. 7	**18.** 30 mo
6. 60	**19.** 18 days
7. 52	**20.** 30 weeks
8. 30	**21.** $\frac{1}{2}$ hr
9. 72	**22.** 60 weeks
10. 150	**23.** =
11. 28	**24.** >
12. 54	**25.** <
13. $2\frac{1}{2}$	**26.** <

27.

28.

29. 3 hr 45 min	**35.** 10 weeks 2 days
30. 18 days 18 hr	**36.** 2 days 13 hr 38 min
31. 6 yr 3 mo	**37.** 20 min
32. 1 min 13 sec	**38.** 15 days
33. 6 min 32 sec	**39.** $2\frac{1}{2}$ min
34. 23 hr 15 min	**40.** 17 hr

48. 6:30 P.M., 10:30 P.M., 2:30 A.M.

49. Every 4 hr

50. 11 A.M., 3 P.M., 7 P.M., 11 P.M., 3 A.M.

51. 2:45 P.M.

52. 200 cu in.

53. 3.375 cu m

54. 9.42 cu ft

55. 560 cu mm

56. 871.87 cu in.

57. 2.37 cu m

58. $3\frac{5}{8}$ in.

59. 35 mm

60.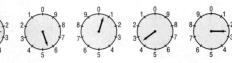

61.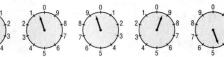

Chapter 6 Pages 128–129

1. $A = 500$ sq ft, $P = 90$ ft

2. 75 sq ft

3. 15 windows

4. 4 feet

5. 100 strips

6. 11 feet 9 in.

7. 2 buckets

8. obtuse

150°

9. complement $= 15°$

supplement $= 105°$

75°

10.

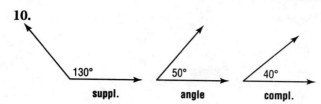

suppl. angle compl.

11. 94 ft

12. 122 ft

13. 420 sq ft (14 × 30)

14. LR 340 sq ft
BR 84 sq ft
K 120 sq ft

15. LR $5\frac{2}{3}$ yd × $6\frac{2}{3}$ yd
K 4 yd × $3\frac{1}{3}$ yd

16. 2 bedrooms and a bathroom

Pages 130–131

1. 3 ft by 3 ft

2. $1\frac{1}{2}$ ft by 3 ft

3. 16 ft by 10 ft

4. $7\frac{1}{2}$ ft

5. 36 ft

6. 100 mi

7. 375 ft

8. 525 m

9. 135 ft

10. 120 km

11. 135 mi

12. 180 mi

Pages 132–133

1. °F

2. 0°F

3. 19°F

4. 10 mph

5. 0°F

6. 35 mph, 40 mph

7. 4°F

8. −12°F

9. −26°F

10. 15 mph wind, 30°F

11. 25 mph wind, 0°F

Pages 134–135

1.

2. 128 lb

3. 29 lb

4. **a.** 31 lb 14 oz
b. about 3 lb 9 oz

5. 49 lb 2 oz

6. 24 lb 9 oz

7. 8 lb 3 oz

8. 3.9 °F above

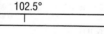

102.5°

9. 3.7 °F below

94.9°

10. 2.7 °F

101.9° 99.2°

11. 5.6 °F above

104.2°

12. 2.5 °F

102.4° 99.9°

13. 1.2 °F

97.4°

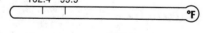

Page 136

1. 12.5 L

2. 5.4 L

3. $18\frac{1}{4}$ pt

4. 25.5 L

5. 4.9 L

6. $61\frac{1}{2}$ qt

7. $32\frac{1}{4}$ qt

8. $9\frac{1}{4}$

  $10\frac{1}{2}$

  10

  $12\frac{1}{4}$

Page 137

1. 2 cans

2. 8 oz

3. 14 c

4. 1 for shortening,
   4 for flour

5. 18

6. $\frac{1}{2}$ c shortening
   2 c flour
   1 tsp salt
   4 tsp baking powder
   2 tsp cinnamon
   1 tsp baking soda
   $\frac{1}{2}$ c sugar
   $1\frac{1}{3}$ c milk
   6 tbsp cocoa
   1 c chopped nuts

7. 8 oz orange juice
   8 oz lemonade
   8 oz strawberries
   16 oz ginger ale
   7 c water or
   $3\frac{1}{2}$ 16-oz cans

Pages 138–139

1. 27,000,000 cu mm

2. $\frac{1}{2}$ cu ft or 864 cu in.

3. 678.24 cu in.

4. 37,200 cu in.

5. 9,300 cu in.

6. 5.23 cu in.

7. 120 cu ft

8. 381.51 cu m

9. $3 \times 3.14 \times 2.5^2 \times 8 = 471$ cu in.

10. 87.92 cu in.

Page 140

1. 4,000 cu ft

2. 14,400 cu ft

3. 8 ft

4. 3,000 sq ft

5. 1,800 cu ft

6. 14,000 cu ft

7. 9,600,000 cu ft

8. 20 ft

Page 141

1. 27 cu in.
   4,212 cu in.

2. 6 cu ft

3. Yes. $36 \times 24 \times 12 = 10,368$ cu in.

4. 35.325 cu in.

5. 14.13 cu ft or 24,416.64 cu in.

6. 49.5 cu ft

7. box 16 cu ft
   leg 0.087 cu ft

8. 47.1 cu ft

Page 142

1. 10 A.M.

2. 11 A.M.

3. 1 P.M.

4. 6:30 P.M.

5. 10 P.M.

6. 3:45 A.M.

7. 12:15 P.M.

8. 10 A.M.

9. 9:40 P.M.

10. 10 P.M.

11. 10:30 A.M.

12. 8 P.M.

13. 10:15 A.M.

14. 2 P.M.

15. 7 A.M.

Page 143

1. 5

2. 1

3. 11

4. 25–31 mph

5. 32–38 mph

6. 8–12 mph

7. Slight damage to buildings, shingles blown off roof

8. Small trees sway; waves break on inland waters

9. Calm; smoke rises vertically

10. Trees uprooted; considerable damage to buildings

11. Widespread damage; very rare occurrence

12. Small branches sway; dust and loose paper blow about

Pages 144–145

1. 20 mpg	11. 780 miles
2. 30.4 mpg	12. 810 miles
3. 19 mpg	13. 180.6 miles
4. 21 mpg	14. 20 mpg
5. 30 mpg	15. 30 mpg
6. 24.3 mpg	16. 1,242 miles
7. 192 miles	17. 434 miles
8. 207.7 miles	18. No.
9. 411.8 miles	19. Yes.
10. 159.96 mi	20. 288 miles

Page 147

1. 6:35	10. 7:17 or 8:02
2. 8:35	11. 7:59 or 8:44
3. 11:01	12. 42 min
4. 6:16	13. 6:16
5. 9:00	14. 6:58
6. 11:44	15. No.
7. 1:26	16. 4:14
8. 4:04	17. 4:40
9. 6:04	18. 7:14 A.M., 6:14 P.M.

Pages 148–149

1. 270 miles

2. 18.68 ≈ 19 hr

3. 63 mph

4. 15.86 ≈ 16 hr + 9 hr on breaks ≈ 25 hours

5. 51 mph

6. 41.6 ≈ 42 hours; 11 P.M. Sunday

7. 10 hours

8. 53 mph

9. 4 hours

10. 6 hours

Pages 150–151

1. $2\frac{1}{2}$ hr

2. $8\frac{1}{4}$ hr

3. 400 mph

4. $\frac{4}{5}$ hr

5. 342 mph

6. 3 hr 15 min

7. 198 mph

8. $5\frac{3}{4}$ hr

9. 460 mph

10. 106 mph

11. 8 hr

12. 402 mph

Posttest p. 152

1. 16

2. 5,000

3. 3.5

4. 47

5. $2\frac{1}{3}$

6. 5,280

7. 3,200

8. 26

9. 0.475

10. 195

11. 365

12. 41

13. $2\frac{3}{4}$ lb

14. $31\frac{7}{8}$ in.

15. $1\frac{1}{4}$ c

16. 84°F

17. 38 min

18. 5 lb 15 oz

19. 3 lb 11 oz

20. 10.4 g or 10,400 mg

21. 33.6 cm

22. 250 km

23. 11 yd 1 ft

24. 2 qt 2 c

25. 26 gal

26. 110 cm

27. 7 days 14 hr

28. 6 days

29. 4 hr 38 min

30. 42 ft 6 in.

31. acute

32. 55°

33. 145°

34. 46 ft or $15\frac{1}{3}$ yd

35. 14 sq yd or 126 sq ft

36. 31.4 ft

37. 78.5 sq ft

38. 39.25 cu ft

39. No.

40. 141.3 lb

41. 161.5 yd

42. Chris

43. 18.2 m

44. 52 g

45. 200.96 cu m

46. 1,000 cu ft

47. $209.30

48. 402.6 miles, 67.1 mph

49. $57.75 for overtime, $673.75 total for the week

50. 128 gal

FORMULAS AND MEASUREMENT UNITS

FORMULAS

Area	
Rectangle	$A = l \times w$
Square	$A = s^3$
Circle	$A = \pi r^2 h$

Volume	
Box	$V = l \times w \times h$
Cube	$V = s^3$
Cylinder	$V = \pi r^2 h$
Cone	$V = \frac{1}{3} \pi r^2 h$

Circumference of a circle

$C = \pi d$ or $C = 2\pi r$

Temperature

$C = \frac{5}{9}(F - 32)$ $F = \frac{9}{5}C + 32$

MEASUREMENT UNITS

Celsius	Temperature	Fahrenheit
°C	water freezes	32°F
100°C	water boils	212°F
37°C	normal body temperature	98.6°F

Time

60 seconds (sec) = 1 minute (min)	24 hr = 1 day
60 min = 1 hour (hr)	7 days = 1 week (wk)
12 months (mo) = 1 yr	52 wk = 1 year (yr)

Metric   *Customary*

Weight

1,000 milligrams (mg) = 1 gram (g)	16 ounces (oz) = 1 pound (lb)
1,000 g = 1 kilogram (kg)	2,000 lb = 1 ton (T)

Length

1,000 millimeters (mm) = 1 meter (m)	12 inches (in.) = 1 foot (ft)
100 centimeters (cm) = 1 m	3 ft = 1 yard (yd)
1,000 m = 1 kilometer (km)	36 in. = 1 yd
	5,280 ft = 1 mile (mi)
	1,760 yd = 1 mi

Capacity

1,000 milliliters (mL) = 1 liter (L)	3 teaspoons (tsp) = 1 tablespoon (tbsp)
1 kiloliter (kL) = 1,000 L	1 fluid ounce (fl oz) = 2 tbsp
	8 fl oz = 1 cup (c)
	2 c = 1 pint (pt)
	2 pt = 1 quart (qt)
	4 qt = 1 gallon (gal)

GLOSSARY

A

angle a figure formed by two lines that start at the same point. The point is called the **vertex** of the angle and the lines are called the **sides**.

Angles can be measured in **degrees**. An **acute angle** is smaller than 90°, a **right angle** is exactly 90°, an **obtuse angle** is between 90° and 180°, and a **straight angle** is exactly 180°.

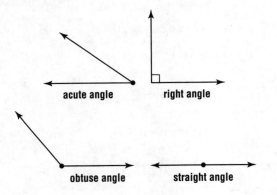

area the physical amount of a region covered by a flat figure. Area can be measured directly, by covering a figure with squares, or area can be calculated based on lengths of parts of the figure.

B

base of a cone or cylinder; the side or sides of a cone or cylinder that are circles

box a container whose top, bottom, and sides are rectangles

C

capacity the physical amount that an object can hold or contain

Celsius a temperature scale. On a Celsius scale, the freezing point of water is 0°C and the boiling point of water is 100°C.

centi- a prefix used with metric units; it means one-hundredth.

circle a round, flat figure. Every point on a circle is the same distance from the **center** of the circle. A line segment from the center to any point on the circle is a **radius**, and a line segment containing two points on the circle and the center of the circle is a **diameter**.

circumference the distance around a circle

compare and order tell whether two objects are the same or different ("compare"), and then tell which one is greater than or less than the other ("order").

complementary angles two angles whose measures add up to 90°

cube a box-shaped object whose six sides are squares

cylinder an object shaped like a tube or a rod; it has circular ends that are the same size.

D

degree a unit for measuring angles; the symbol is °. There are 360 degrees in a full circle. (It is also a unit for measuring heat.)

diagonal a line segment that connects two corners of a figure that are separated by at least one other corner of the figure

dimensions a term used to refer to the length and width of a rectangle or to the length, width, and height of a box

distance the physical amount of length between two points or objects

duration another term for "amount of time" or "time interval"

E

equivalent measurements measurements that represent the same amount. For example, the three measurements 36 inches, 3 feet, and 1 yard are equivalent.

estimate as an object or noun: an approximate value; as an action or verb: to calculate or make an educated guess for an approximate value.

F

Fahrenheit a temperature scale. On a Fahrenheit scale, the freezing point of water is 32°F and the boiling point of water is 212°F.

H

height a physical amount of distance, measured vertically or up-and-down

K

kilo- a prefix used with metric units; it means one thousand.

L

length a physical amount of distance

level a carpenter's term for horizontal

M

measurement a number and a unit. Also, as a general term *measurement* means using tools to assign numbers and units to objects, and then using those numbers and units to compare and relate objects.

milli- a prefix used with metric units; it means one-thousandth.

O

operation a term used to refer to addition, subtraction, multiplication, or division

P

parallel lines two or more lines that go in the same direction and never meet. When two parallel lines are crossed by a third line, they form four pairs of **corresponding angles** and four pairs of **alternate interior angles**.

perimeter the distance around a figure

plumb a carpenter's term for vertical or straight up-and-down

protractor a tool used for measuring angles. It usually has numbers 0 through 180 marked along a half-circle.

R

rate another term for velocity or speed. In general, rate can refer to any fraction where the numerator is a measurement and the denominator is a measurement that uses the number 1.

ruler a tool for measuring length

S

scale any tool for measuring, especially for measuring weight

scale drawing a drawing of an object where the lengths in the drawing represent lengths on the actual object

scale on a map a key telling how the lengths on the map represent actual distances

semicircle half a circle

sphere a ball-shaped object

supplementary angles two angles whose sum is 180°

T

temperature the physical feeling of hot and cold

thermometer a tool for measuring temperature

thermostat a device for regulating temperature. It uses a thermometer to tell when to turn a heating or cooling system on and off.

time measured in seconds, minutes, hours, and so on. The tool used for measuring a *duration of time* is the stopwatch; the tool used for recording time is the watch or clock.

time zone one of 24 regions of the Earth; as you cross a time zone, you change the time showing on your clocks by one hour.

U

unit the amount of length, capacity, or velocity that represents one inch, one cup, one mile per hour, and so on. **Customary units** include inches, pounds, and quarts; **metric units** include meters, grams, and liters.

V

velocity how fast an object is traveling. Usually, velocity is not measured directly, but is calculated based on traveling a particular distance in a particular time.

volume the physical amount of space used by an object. Usually, volume is not measured directly but is calculated based on lengths of parts of the object.

W

weight the physical feeling of how heavy or light an object is. The tool for measuring weight is called a scale.

Z

zero mark the starting place on a ruler or scale.

INDEX